FEEL-GOOD HOMES

HOW TO CHOOSE THE RIGHT
HEAT PUMP FOR A COMFORTABLE,
HEALTHY, SUSTAINABLE HOME

DREW TOZER

FOREWORD BY NATE ADAMS

Re^think

First published in Great Britain in 2025
by Rethink Press (www.rethinkpress.com)

© Copyright Andrew Tozer

All rights reserved. No part of this publication may be reproduced, stored in or introduced into a retrieval system, or transmitted, in any form, or by any means (electronic, mechanical, photocopying, recording or otherwise) without the prior written permission of the publisher.

The right of Andrew Tozer to be identified as the author of this work has been asserted by him in accordance with the Copyright, Designs and Patents Act 1988.

This book is sold subject to the condition that it shall not, by way of trade or otherwise, be lent, resold, hired out, or otherwise circulated without the publisher's prior consent in any form of binding or cover other than that in which it is published and without a similar condition including this condition being imposed on the subsequent purchaser.

Cover photograph by Mat Krizmanich

To Amanda, Momo, and Lenny, who inspired the journey to create a feel-good home that's filled with love; and to every homeowner looking to do the same.

Contents

Foreword   1

Introduction: What Is A Feel-Good Home?   3
   The great indoors   5
   The underlying issue   9
   The way forward   11

PART ONE   Where We Are Now   13

1   The Current State Of An Average House   15
   How good is your house?   16
   Is your furnace too big?   18
   Is it drafty in here?   24
   Something in the air   28
   How wet is *too wet*?   38
   Summary   41

2 The Difference Between Good And Bad Contractors — 43

Not all contractors are suited to help — 44
Getting the calculation right — 55
Summary — 59

3 The Future Of Heating And Cooling — 61

Heat pumps 101: How do they work? — 62
Fossil fuels aren't future-proof — 66
Preparing for climate change — 68
Summary — 71

4 The Path To A Feel-Good Home — 73

The universal benefits of right-sized HVAC — 74
Why right-sized means heat pumps — 79
Less (maintenance) is more (value) — 81
Summary — 82

PART TWO   Finding The Right Path — 83

5 Introduction To The HAVEN Method — 85

The steps to creating a HAVEN — 86
The shortcut for simple homes — 87
Summary — 88

6	**Step 1: Assess The House**	**91**
	H—Heat load	93
	A—Air leakage	99
	Summary	103
7	**Step 2: The Right Mindset**	**105**
	V—Value mindset	106
	Summary	110
8	**Step 3: Create A Plan**	**113**
	E—Environmental control	113
	N—Necessary infrastructure	119
	Summary	125
9	**Step 4: Implement—Do The Work**	**127**
	Execute the plan	127
	Embrace each step	128
	My house, before and after	129
	Summary	138
10	**"But The Grid!" (And Other Concerns)**	**139**
	Cold weather, electricity generation, and the grid	140
	Ground-source heat pumps	145

Replacing other gas appliances	147
Building envelope and window upgrades	150
Conclusion: Your Feel-Good Home	**153**
Spark a feel-good community	156
Live in a feel-good home	157
Notes	**159**
Acknowledgments	**171**
The Author	**173**

Foreword

I've known Drew since 2020. I'm happy that he wrote this book, and I'm even happier that you're holding it. Drew is funny, knows his stuff, and is a realist about getting things done. That's an exceedingly rare combination.

My quest for years has been making it so that anyone can get a comfortable and healthy home, or as Drew calls it, a "feel-good home." This book is part of that quest. Drew has been a great protégé and balances what needs to be done with the art of the possible.

Rather than send you alone on a quest for perfection that's not likely to succeed, Drew and his book will guide you through simple steps to get a feel-good home with methods that can easily be done by good local contractors (or by Drew himself, if you're lucky enough to be in his area!).

Grab your favourite beverage, sit in your favourite chair, and set aside your preconceived notions about HVAC and home comfort. Read this book with an open mind, and I can almost guarantee that the next time you grab that beverage and sit in that chair, it'll be far more comfortable. If you're like any of our clients, you may realize you don't like being out of the house as much. It's just nicer to be in a feel-good home.

Good luck!

Nate "The House Whisperer" Adams
Author of *The Home Comfort Book*

Introduction: What Is A Feel-Good Home?

Feel-good homes offer a combination of remarkable home comfort, clean air, resilience, sustainability, and low maintenance. They are remarkable places to live.

The average house presents a range of problems for its occupants: rooms that are too hot or too cold; moisture issues; large temperature swings between floors; asthma, allergies, and respiratory problems from poor indoor air quality; and high-maintenance systems that are ignored until they break. These are all solvable problems, and good contractors are the experts that can help. You just need to find contractors who care enough to assess, plan, and implement the right solutions.

We'll discuss the concepts of indoor environmental quality (IEQ) throughout the book. IEQ is a metric of indoor conditions—it includes everything that impacts how you feel in a space. While IEQ incorporates factors like acoustics and lighting, our focus in this book will be on thermal comfort and air quality.

I have created the HAVEN method as a step-by-step path for homeowners and contractors to follow. It's a process to control indoor conditions and transform average houses into feel-good homes.

HAVEN stands for:

- **H**—heat load
- **A**—air leakage
- **V**—value mindset
- **E**—environmental control
- **N**—necessary infrastructure

INTRODUCTION: WHAT IS A FEEL-GOOD HOME?

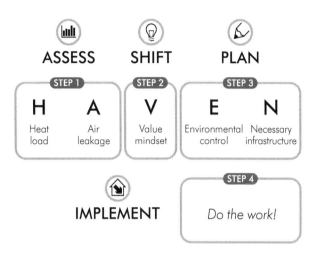

The HAVEN method

To start, we have to identify the needs of the house and work with the homeowner's goals and budget to find the right project to solve meaningful problems. Part One of this book explains the challenges and opportunities involved, and Part Two will outline my HAVEN method, which will help you transform your own home.

The great indoors

On average, we spend 68.7% of our time indoors at home.[1] That's 250 days (more than eight months) each year, or twenty-one days per month. If we live to eighty, it equates to fifty-five years spent inside our homes. Creating a feel-good home means spending those fifty-five years in a remarkable place.

Most homes, however, aren't feel-good homes. Most homes barely feel OK. We've accepted a narrative that *It is what it is* when it comes to indoor spaces. People reject the *as-is* mentality in most areas of their lives; we demand high performance from electronics, internet speeds, cell reception, vehicle safety, professional services, and restaurants, and the list goes on. Despite spending most of our lives indoors, though, we accept mediocrity in home performance.

I created the HAVEN method to show you the path to a better home. Your house can be a feel-good home, regardless of its size, shape, or location. If you follow the HAVEN method, you'll end up in a home with fresh, clean air that is a haven from allergens, pollutants, and airborne diseases. It will be remarkably comfortable, and low carbon to align with a sustainable future.

In some homes, finding and installing a right-sized heat pump can provide a quick fix. Other houses require a longer process, with air sealing or other energy upgrades to help the house reach a tipping point at which the indoor conditions can be confidently controlled.

Every house and every homeowner is unique, so the first steps depend on your needs, goals, and budget. It's a personal experience that requires honesty about *where you are* and *where you're going*.

Hey, narrator?

Oh, hey—that *is* personal. Hi, reader.

Why should I listen to you?

Good question, and it's a great segue to an introduction. My name is Drew Tozer. I'm a founding partner of Foundry Heat Pumps. By trade, I'm an energy advisor, with expertise in building science (the science of understanding how heat, air, and moisture flow through buildings) and heat pumps. I developed the HAVEN method in the field while helping homeowners solve problems that mattered to them.

In 2022 I electrified my 1920s double-brick house in Port Hope, Ontario, Canada. This solved some comfort problems I was having on a previously unusable third floor, reduced hot water wait times and costs by 50%, improved indoor air quality, and reduced carbon emissions by 95%.

It had been a quintessential *as-is* house. Previous homeowners had accepted it as drafty and uncomfortable for decades. They had a fixed mindset about its potential. I used the HAVEN method to turn it into a feel-good home.

As an energy advisor with boots on the ground, I realized that advisors and contractors are incentivized to rush through houses and deliver bare minimum results—more houses mean more revenue and profit. Heating, ventilation, and air conditioning (HVAC) contractors have an especially low bar for what they think is *good customer service*.

The reality is that nobody knows your house like you do. It takes time and effort to understand how it operates, and that process doesn't fit into traditional contractor business models. A contractor can't diagnose problems in your house during a fifteen-minute sales call.

I believe that everyone deserves to live in a feel-good home, with an abundance of comfort and clean air. It's not realistic to expect homeowners to complete that journey by themselves, though. HAVEN is a path that lets contractors thrive while solving homeowners' problems and reducing emissions.

Hmm, I believe you.

Great! Let's talk about everything that's wrong with your house.

How do you know something's wrong with my house?

You're reading the book, aren't you? My hope is that it's a small problem with a simple solution, so you can quickly live in a feel-good home—but even if it's a complicated house, we'll follow the same process and reach the same remarkable result, it will just take a bit longer.

The underlying issue

In the 1980s, we started seeing issues related to *sick building syndrome*, which was a catch-all term for illnesses caused primarily by living or working in buildings with poor indoor air quality (IAQ).[2] Problems were made worse when we embraced air sealing as an energy efficiency tool without considering its impact on airflow, mould growth, and ventilation rates.

In modern building science, we have a concept known as *house as a system* (HAAS) to avoid such unintended consequences. HAAS is the idea that every part of a house is interconnected, so a small change in one area can significantly impact others.

There is a positive side to this relationship, though: solutions are often interconnected. If you successfully solve a temperature problem by installing a right-sized heat pump with proper specifications, that solution could improve indoor air quality and humidity control while reducing carbon emissions.

Most homeowner complaints can be solved by effectively controlling the indoor environment. Problems and solutions can be complicated, though, and complicated solutions can be expensive.

It sounds like a pain. Why would anyone do it?

The payoff is enormous, and we're aiming for remarkable results. Once you live in a feel-good home, you'll be annoyed whenever you're in a different home. Everything will feel a bit *off*. The ultimate goal is to find and solve underlying problems to create better homeowner experiences.

I've written this book so you can skip the most painful parts of the process: starting from scratch with no direction and limited understanding of how your house works. This book introduces some scientific insights to help you understand the scale of the problems and solutions. Plus, building science facts are great at parties!

I doubt that.

Yeah, you're right, but a basic understanding of building science will add important context. I'll only include the most valuable science lessons. This isn't a textbook.

The way forward

The HAVEN method puts homeowners on the path to feel-good homes. It encourages better conversations between contractors and homeowners, and it helps to find the right heat pump for each house.

The overarching theme is clear: your house isn't performing at its best, and it has the potential for more. There are solvable problems in your house, and you're looking for a way to fix them. While you're reading this book, try to keep *your* house and its problems in mind so you can work out how the HAVEN method best applies to you.

PART ONE
WHERE WE ARE NOW

ONE
The Current State Of An Average House

A house serves an important role in providing comfort, health, and safety, but those factors are ignored during renovations and upgrades. In this chapter, we'll discuss the imperfections of home performance and its impact on day-to-day life.

From oversized HVAC systems to indoor air quality and managing humidity and water, the first step to solving problems is to investigate the underlying issues.

How good is your house?

Before we go into the details of the HAVEN method, let's do a quick assessment to see how much help your house needs. These questions are designed to give you an understanding of the current state of your house. I've included prompts to guide your thinking.

Rate your home from 1 to 10 in each category, where 1 means *Not a problem* and 10 means *It's a serious problem that needs to be fixed ASAP*.

Category 1: Home comfort

Rank from 1 to 10: My house has comfort problems in the summer.

Rank from 1 to 10: My house has comfort problems in the winter.

Prompts:

- Do you have a room or floor (eg the top floor) that's too hot in the summer or too cold in the winter?
- Do you have central AC but need to use a window AC or fans for extra cooling?

- Is your bedroom uncomfortably hot or cold at night?

- Do you use a space heater to make one room of the house more comfortable in the winter (eg in a home office)?

- Do you close doors to isolate rooms or floors because they're too hot/cold?

Category 2: Health and air quality

Rank from 1 to 10: My house makes respiratory symptoms worse.

Rank from 1 to 10: My house has health risks related to poor indoor air quality.

Rank from 1 to 10: My house has humidity problems.

Prompts:

- Does anyone in the house have asthma or respiratory issues?

- Does anyone in the house struggle with seasonal allergies?

- When one person in the house gets sick, does everyone usually get sick?

- Have you noticed mould in the house?
- Is any room (eg your basement) too humid in the summer?
- Do you have standalone dehumidifiers to try to control humidity levels?

It's important to keep *your* house in mind as you read this book. Your problems are the ones we're trying to solve, and high scores are a sign that there are problems worth solving.

Is your furnace too big?

An oversized furnace is bad for comfort and air quality, and 95% of furnaces I've come across in the field are oversized. Many are two to four times too big.

In older homes, furnaces have been oversized for decades, with every emergency replacement being like-for-like—a dying oversized furnace swapped for a newer model in the same (over)size.

Unsurprisingly, that leads to the same comfort and air quality problems in the house. All too often, heating systems are seen as basic utilities rather than a tool to solve homeowner problems. That's a mistake. Installing a right-sized heat pump is the most important step to creating a feel-good home.

Note: In this book I refer to *gas furnaces,* but the ideas and lessons apply to all types of combustion (fuel-burning) furnaces. If you have, for example, a propane furnace, simply replace *gas furnace* with *propane furnace* in your mind as you read.

SCIENCE LESSON #1:
What is a heat pump?

Heat pumps are a type of air conditioner that can both heat and cool.

That's my one-liner for homeowners.

Is it that simple?

Yes, pretty much. Heat pumps use the same process as air conditioners (called the *refrigerant cycle*). Heat pumps just have an extra part: the reversing valve.[3]

In the summer, heat pumps act like modern air conditioners (AC).

In the winter, the reversing valve lets the heat pump run in the other direction. It heats the house instead of cooling it.

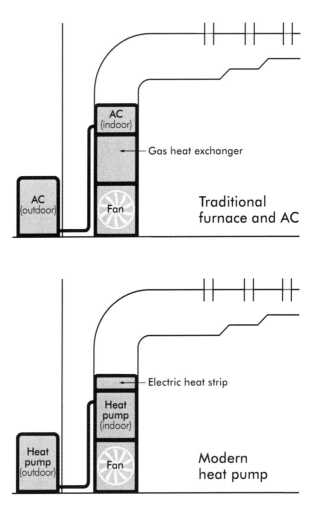

Furnace and AC versus heat pump

THE CURRENT STATE OF AN AVERAGE HOUSE

It's common for homeowners to think that ACs "add cold" to a house, but that's not technically right. Picture a lightbulb in a windowless room. You don't make the room dark by "adding darkness"—you make it dark by turning off the light.

Similarly, ACs don't add cold—*they remove heat*. Heat pumps and air conditioners move heat from one space to another (from inside to outside or vice versa).

A heat pump includes both an indoor and outdoor unit. The indoor air handler (fan) looks nearly identical to a furnace, and the outdoor unit looks like a traditional AC.

It can replace both the furnace and air conditioner in a house, or it can replace the AC and work with an existing furnace.

Whose fault is it that a home's furnace is too big?

Everyone blames contractors for oversizing equipment, but it's partially homeowners' fault. Most homeowners pick HVAC solely on price, without thinking about the bigger picture or using it as an opportunity to improve the house. They pick the lowest quote and get the cheapest solution. The contractor joins this *race to the bottom* to try to win jobs based on price.

FEEL-GOOD HOMES

Is it the contractor's fault for joining the race or the homeowner's fault for asking them to? I say both. Fault falls into the interaction between homeowners' priorities and contractors' business models.

Here's a thought experiment to explain another reason for oversized HVAC. Imagine a world where people are as clueless about cars as they are about HVAC. They don't know the difference between makes and models, and nobody cares what type of car anyone drives. There's no such thing as a *cool car*.

In that world, you happen to own a minivan, which you use every day to commute to work. When it breaks down and you need a replacement, the salesperson is quick to recommend an affordable, three-row minivan with seven seats and plenty of storage. They don't ask about your needs, how you'll use it, or how many people you drive around.

This is the salesperson's thought process:

1. A van is big enough for most families (so it's probably big enough for you).

2. It's flexible enough to do most things (so it will probably do what you want).

3. Your last vehicle was a van (so it's probably what you're expecting).

You go to a second car dealership and they recommend a similar van. You need to make a decision right away because you need it for commuting. The van is in stock, so you buy it. It turns out, though, that a smaller car would have been a better fit. The van is too big for your needs—you rarely have more than two people in it. It's oversized.

Why didn't the salesperson recommend a smaller car? It's because you were looking for a quick and easy replacement. They picked a one-size-fits-all option—the lowest common denominator that's *good enough* for most people.

This is similar to the HVAC sales process. Contractors recommend basic, oversized furnaces instead of spending time to find the right option for each house. This reduces the contractor's upfront work, and it reduces the chance that a homeowner gets a furnace that's too small.

Contractors don't take the time to align your goals, needs, and budget for the best solution. Undersized equipment leads to callbacks, which cost the company money and hurt their bottom line. They therefore take the easy way out, installing oversized equipment, and homeowners don't have enough information or context to push back.

What's the right size for me then?

I love that question.

Thanks! I thought of it myself.

I wish there was an easy answer, but every house is different. Discovering the right size requires a *heat load calculation* to determine how much heating your house needs to stay warm on a cold winter day.[4] It's a relatively simple calculation, but most HVAC companies skip it.

Is it drafty in here?

Air leakage plays a key role in comfort, humidity, and energy efficiency. In leaky houses, air sealing is required to create a feel-good home. Forget what you've heard: old houses don't need to be drafty.

Sounds like an easy fix.

In theory, yes, but every hole in the building envelope (walls, ceiling, windows, doors, etc) is a path for air to leak in and out of the house. The best time to air seal is during construction. The second-best time is today. The third-best time is during your next renovation.

How leaky is my house?

The only way to know is to test it.

A *blower door test* uses a big fan set in a doorway to blow at different speeds and measure the pressure difference between inside and outside. That gives us real data on the leakiness of the house, but sealing holes isn't as simple as spotting a gap under a door. Rather, it's a three-dimensional maze behind the walls, where air enters the house in one area and exits via another. Air just needs a path, it doesn't care if it's straight or meandering.

Homeowners have for decades been told that houses need to breathe. It's a misconception that makes homeowners feel powerless in old, drafty, uncomfortable houses. It leads to the *It is what it is* mentality.

The science is clear, though: a house does *not* need to breathe. It does *not* need to be drafty. Building materials need to *dry*, and homeowners need *fresh air*. It seems like a small distinction, but it can be the difference between an old and drafty house and a feel-good home.

The goal isn't to find *all* the holes. It's to make the house *good enough* so that you can control the indoor conditions with a right-sized heat pump.

SCIENCE LESSON #2:
The stack effect

For air to move, it just needs a pressure difference and a path. Air naturally moves from high to low pressure.

Outdoor air pressure decreases with altitude (height), but it is relatively constant within a house. At the top of the house (higher altitude, lower outdoor pressure), the outdoor pressure is *less* than the indoor pressure, so air moves *out* of the attic.

Conversely, at the bottom of the house (lower altitude, higher outdoor pressure), the outdoor pressure is *greater* than the indoor pressure, so the air moves *into* the basement.

The overall effect is that air leaks into the basement, travels up through the house, and out the attic or ceiling. The house acts like a chimney stack, so this phenomenon is aptly called the stack effect, and it plays an important role in air leakage and home performance.[5]

The stack effect increases with building height because taller buildings have bigger pressure differences at the top and bottom. Fun fact: this is the origin of the revolving door—the stack effect was so strong in tall office buildings that lobby doors were difficult to open and caused a rush of cold air into the lobby in the winter. Revolving doors let people come and go without opening a hole in the building envelope.[6]

You can use the stack effect to your advantage. Prioritize sealing holes near the top (attic) and bottom (basement) because they leak considerably more than holes near the middle of the house.

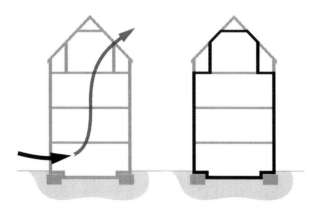

Stack effect and building envelope

Something in the air

As I mentioned in the introduction to this book, the average Canadian spends 68.7% of their time indoors at home. When you turn eighty, you will have spent fifty-five years inside your house.

I'm only sixty-two, though!

OK, no need to brag, you've still spent forty-three years indoors already, more or less. These are averages, but at any age, you've been indoors for most of your life.

As much as we dislike some smells, it's more dangerous when pollutants are odourless because they're easier to ignore. Unfortunately, many pollutants—including carbon monoxide, radon, CO_2, fine particulate matter (PM2.5), allergens, and airborne viruses and bacteria—are odourless.

Regardless of odour, there are only two options for dealing with air pollution:

1. Measure pollutants and mechanically clean the air
2. Allow your lungs to act as the filters

THE CURRENT STATE OF AN AVERAGE HOUSE

The unfortunate reality is that most people default to the second choice. They let their lungs filter pollutants and deal with the resulting health and wellness impacts.

It's relatively easy to make longest-lasting improvements to IAQ because homeowners have direct control over the mechanical equipment in their house. In most cases, though, it's low on a long list of priorities.

By contrast, we have high standards for water quality. Water goes through strict filtration and testing to ensure it's clean. We *expect* clean water. Nobody drinks dirty water and thinks *It's fine—my liver will filter the pollutants*.

We're at the edge of a paradigm shift in realizing the importance of clean air in our communities. The public's growing interest in air quality is a lasting legacy of the COVID pandemic.

A similar shift happened when cholera and other waterborne diseases changed how we test, monitor, and treat water. It was recognized that improving water quality was required for public health. We need to think the same way about clean air.

After all, you deserve to live in a house with clean air.

How do you know the air in my house isn't clean?

It's a fair question. I don't know if it's clean, but I can almost guarantee that you don't know if it's clean either. You can't know unless you're monitoring it.

There have been significant advancements in IAQ monitoring over the last decade, with higher-quality sensors finding their way into the residential market, but the list of potential air pollutants is overwhelming. For that reason, I recommend focusing on two pollutants in particular: CO_2 (carbon dioxide, a gas) and PM2.5 (tiny particulates that are 2.5 microns or smaller).

Why CO_2 and PM2.5?

Monitoring PM2.5 and CO_2 gives valuable feedback about the effectiveness of filtration and ventilation systems. If you have the means, ability, and interest to monitor additional pollutants, there's value in knowing what else is in the air—but choosing PM2.5 and CO_2 is an effective and accessible starting point that's good enough for most homeowners.[7]

Humans are the main source of CO_2 in buildings (we generate it when we breathe), so monitoring CO_2 levels is a useful tool to check ventilation rates. High levels mean CO_2 is accumulating, which means poor ventilation. For the same reason, CO_2 levels are used as a proxy for the risk of COVID and other airborne diseases. Ventilation rates are only part of the risk equation, but it's a simple measurement that provides useful insights.[8]

PM2.5 is common in homes, where it's produced by combustion, cooking, smoking, and cleaning.[9] Monitoring PM2.5 levels tells you about filtration performance—if PM2.5 levels are high for an extended period, there's a greater risk of airborne viruses and bacteria because those similarly-sized particles are likely not being removed by the filters either. Conversely, consistently *low* PM2.5 levels reflect a well-functioning filtration system. PM2.5 also leads to long-term respiratory and other health issues by entering the bloodstream through the lungs, so there's a direct health risk associated with elevated PM2.5 levels.[10]

We've mentioned mechanical solutions several times in this section, and while monitoring air pollutants is a good first step, action is required to improve IAQ by removing them from the home. The three common types of mechanical solution are:

- Source control
- Filtration
- Ventilation

Source control

There aren't many famous quotes about indoor air quality (IAQ), but Joseph Lstiburek attributes the following to the chemist Max von Pettenkofer: "If there is a pile of manure in a space, do not try to remove the odour by ventilation. Remove the pile of manure."[11]

Attempting to remove manure odour with a fan

The lesson in the manure story is simple: to achieve better air quality, you don't start with ventilation or filtration, you remove the source of pollutants.

Source control measures can be *proactive* or *reactive*. *Proactive source control* means choosing products and appliances that don't cause IAQ problems. Buying nontoxic cleaning supplies, installing an induction stove, or quitting indoor smoking are examples of proactive source control.

Reactive source control means removing pollutants when and where they're generated, including kitchen exhaust hoods (removing pollutants from cooking) and bathroom fans (pollutants from toilets and humidity from showers).

Filtration

Homes with central HVAC systems have a furnace filter, but it's usually undersized and low efficiency. HVAC contractors and manufacturers aren't in the business of clean air. It's a misconception that the main role of a furnace filter is to protect the equipment from dust and debris.

Thick, pleated high-efficiency media filters improve IAQ by capturing particles without negatively impacting the HVAC system.

There are different types of electronic air cleaners, including *ionization*, *photocatalytic oxidation (PCO)*, and *ozone generators*.[12] Electronic air cleaners are a common upsell from contractors but they should be avoided because they're ineffective and potentially dangerous—they promise to *neutralize* pollutants by adding chemicals to the air, but the resulting science experiment is unpredictable.

HEPA filters are unnecessary for most houses—they are better suited for high-risk settings like hospitals. They are effective but expensive and MERV 13 filters with high airflow often outperform expensive HEPA systems anyway.[13]

Regardless of the solution, it's important to test duct pressure before and after central filtration is installed to ensure long-term durability of the HVAC equipment. Houses without central HVAC often lack filtration and ventilation solutions.

Ventilation

Ventilation helps by adding "fresh" outdoor air to dilute indoor pollutants, including gas. Ventilation is only helpful if the outdoor air is clean, which isn't true during wildfires, smog days, or at other times when the outdoor air quality is poor.

There are three types of ventilation: supply-only, exhaust-only, and balanced. My preference is *supply-only ventilation*, because it's affordable, effective, and almost zero maintenance. It's simply a fresh air duct connected to the existing furnace or heat pump.[14]

Exhaust-only ventilation isn't recommended as a whole-house solution, but balanced ventilation (heat or energy recovery ventilators) is an effective tool to improve IAQ, especially in high-performance homes.

Homeowners should focus on simple, effective, low maintenance options: remove toxic chemicals, don't smoke indoors, use kitchen and bathroom fans, and install a right-sized heat pump with a MERV 13 filter and fresh air duct.

SCIENCE LESSON #3:
Condensation and dew points

Everyone is familiar with condensation. It's the water that appears on the side of a cold glass on a hot summer day. It's also common to find condensation on old, single-pane windows in the winter.

Where does the water come from?

Air contains water vapour, and air's capacity to hold water increases with temperature, meaning warmer air can hold more water than colder air.

Condensation is the process whereby humidity (water vapour) is forced out of the air and becomes liquid as the air temperature drops, like a shrinking glass overflowing with water.

THE CURRENT STATE OF AN AVERAGE HOUSE

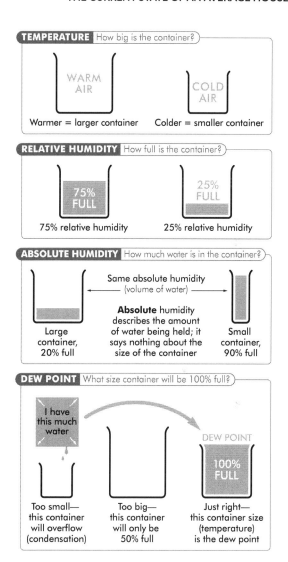

Understanding condensation

How wet is *too wet*?

One of a building's main jobs is to keep water out. Bad things happen when buildings get wet, but bulk water infiltration (like a roof leak) isn't the only concern.

Humidity and condensation can lead to equally bad results. Humidity issues are usually caused by air leakage and made worse with oversized or poorly performing HVAC systems.

The ideal indoor humidity range changes throughout the year, but it's reasonable to aim for 30–40% in the winter and 40–50% in the summer. Humidity targets outside of those ranges can lead to wasted energy, increased costs, or poor IAQ.

Low humidity in the winter is frequently tackled with a whole-home humidifier—a band-aid solution because air leakage is the problem. The humidifier will keep the house comfortable, but there's a risk of condensation and water damage if the humid air leaks into the walls and condenses on cold building materials. Houses that struggle with low winter humidity often don't need ventilation because they're already too leaky.[15]

The problem is flipped in the summer, when we're trying to keep humidity levels from getting too high. Humidity plays a key role in comfort—high humidity makes it more difficult for the human body to cool itself off by sweating. Homeowners often lower thermostat temperatures to combat high humidity, but that can lead to over-cooled, equally uncomfortable houses.

Would buying a new AC fix that?

Maybe, or it could make it worse.

There's a push for higher-efficiency equipment. It's logical to maximize cooling output to energy input, but that comes at the expense of dehumidification. This includes heat pumps operating in cooling mode.

ACs and heat pumps remove two types of heat: *sensible heat* (the temperature that you feel) and *latent heat* (the humidity). How much of each gets removed depends on the equipment, and the relationship is measured by the *sensible heat ratio* (SHR). The easiest way for manufacturers to increase efficiency is to focus on sensible heat removal. That means prioritizing temperature reduction over dehumidification. In practice, equipment with higher efficiency often results in worse dehumidification performance.

Is that a problem?

If you live in Arizona, where it's hot but dry, then no. The higher SHR aligns with your needs: you have a lot of sensible heat to remove, with no latent heat.

If you live somewhere with higher humidity (including Ontario), it could be a problem. We set our thermostats based on temperatures, so your AC runs until the sensible heat is removed … and then it stops.

Do right-sized heat pumps help with any of this?

Oversizing equipment exacerbates the problem because shorter runtimes result in less dehumidification. During shoulder seasons—spring and fall—you won't get any dehumidification at all. The only way to remove the excess humidity is by over-cooling the house or buying an additional dehumidifier.

The best solution for most homes is a right-sized heat pump with longer runtimes, good dehumidification performance, and the option for *reheat dehumidification*. This is an uncommon option right now, but it will likely be added to more models in the future. Reheat dehumidification allows heat pumps to act like whole-home dehumidifiers.

In Canada, heat pumps are sized for heat loads, but smaller units perform better for cooling and dehumidification, reducing significant health hazards related to moisture problems. Cooling and dehumidification loads are important considerations for contractors during the equipment selection process.

Summary

- It's important to keep *your house* in mind as you read this book. Your problems are the ones we're trying to solve.

- Most houses have furnaces and air conditioners that are too big.

- Oversizing equipment is the underlying cause of many homeowner complaints, including comfort, air quality, and moisture problems.

- Air leakage is the biggest cause of heat loss in houses. The easiest time to do air sealing is during construction, but it's often overlooked at that stage. Air sealing can be completed as a separate project.

- We spend nearly 70% of our time in our homes, and our health depends largely on the quality of the air that we breathe during that time.

- The first step to good IAQ is measuring it. At a minimum, CO_2 and PM2.5 should be monitored to confirm that filtration and ventilation systems are operating effectively.

- Controlling indoor humidity is important for comfort and the long-term durability of building materials. Air leakage is often the underlying cause of humidity issues, while modern, high-efficiency HVAC can make it worse.

TWO
The Difference Between Good And Bad Contractors

The quality of home renovations and energy upgrades depends primarily on the quality of the contractor doing the work. When I was an energy advisor and consultant, I provided unbiased direction to homeowners, but successful projects require good contractors. Traditional HVAC companies target *just good enough* with outdated ideas and oversized equipment.

This chapter looks at contractors' role in the process, highlights red flags to watch out for, and reveals a path to finding better local contractors that can help.

Not all contractors are suited to help

I love the HVAC industry. I do.

But...?

HVAC is a powerful tool that can be used to solve many homeowners' problems and create feel-good homes, but contractors don't use it that way. Contractors market themselves as *comfort experts* and *comfort advisors*, but most of their business is emergency equipment replacement. They make sure houses are warm in the winter and cool in the summer, and that's where the problem-solving stops.

Emergency replacements can be avoided by proactive planning, but many homeowners find it difficult to proactively replace equipment for multiple reasons—primarily time, budget, and effort constraints.

Why don't HVAC contractors help more?

It's not what traditional HVAC businesses are designed to do.

DIFFERENCE BETWEEN GOOD AND BAD CONTRACTORS

Entrepreneur Daniel Priestley talks about a business framework with four key skillsets that combine to drive success for companies.[16] He pairs them to the four suits in a deck of cards:

- **Clubs:** "Head in the clouds," CEO—responsible for big-picture, long-term, visionary planning. *What is the big problem in the world that we're trying to solve?*

- **Spades:** "Doing the work," COO—responsible for day-to-day operations. *How do we successfully deliver the product or service to customers?*

- **Hearts:** "Connecting with people," CMO—responsible for making sure the company connects with customers and partners to deliver value. *Who is the ideal customer, what is their current versus desired situation, and what obstacles are stopping them from getting there?*

- **Diamonds:** "Money, finance, data," CFO/CTO—responsible for creating a product ecosystem that leads to a profitable company. *How can we offer a group of products to make the customer journey and decision-making process easier, while creating a profitable company?*

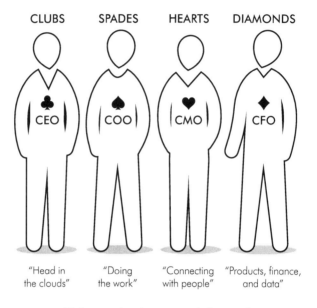

Clubs, spades, hearts, and diamonds

The traditional origin story for an HVAC company goes like this: a skilled tradesperson gets hired as an apprentice for another contractor. They work their way up and become a top-performing technician (*tech*). Eventually, they leave to start their own company, usually for reasons like *The boss is making money off my work*, or *I don't like the choices the owner makes*, or *I do more work than the owner*.

DIFFERENCE BETWEEN GOOD AND BAD CONTRACTORS

That happens again and again. A tech joins a company and becomes skilled enough that they'd rather run their own company than work for someone else. That's why most HVAC companies are small (fewer than ten employees) and local. Those techs are *spades*—they're incredibly skilled at doing the work—so HVAC companies are built entirely around that strength with no focus on the other skillsets (big-picture ideas, connecting with people, and well-designed products).

In many ways, homeowners reward HVAC companies for this business model.

So it's my fault?

It's not *not* your fault.

Compare the HVAC contractor to a local restaurant that wants to serve delicious, cultural food to happy customers and make a profit. What would happen, though, if customers showed up and kept ordering generic, value meals? What happens if every customer sends the signal that they want cheaper, faster food? What should the restaurant do?

Sell value meals!

Exactly. That's what the people want, so that's what the restaurant starts to sell. Soon, though, every restaurant in the neighbourhood is selling value meals because they're all getting the same signal from customers. The restaurants can't differentiate their products because they'll lose customers to the restaurant next door.

This is called *commoditization*. It's when a product or service in an industry becomes so standardized that the only differentiator is price. Food, fast fashion, and airline tickets are commoditized products. We view them as interchangeable, so we choose the lowest price.

Interestingly, labour is also commoditized. Picture two landscaping companies, let's call them Premium Yards and Budget Grass Works. A homeowner gets quotes from both companies, and Premium Yards is 20% more expensive. Premium Yards uses skilled labour, with more experience and lower employee turnover because of the higher pay. They do higher-quality work, with thoughtful recommendations and add-ons based on their experience. Budget Grass Works is the opposite: they pay minimum wage, so they're constantly hiring and training new workers with minimal experience. However, landscaping, like most home services, is commoditized, Homeowners don't understand the difference in skill levels, so they pick the cheaper quote. If that happens enough times, Premium Yards is forced to lower their prices by using less experienced workers to stay competitive.

Homeowners treat HVAC as a commoditized product, with price as the deciding factor. Value isn't given to skilled, experienced labour or high-quality equipment. It's not anyone's fault—it's a chicken-and-egg problem. Contractors realize that homeowners select companies based on price, so they continue to cut costs by using less-experienced labour and cheaper equipment. Homeowners continue to get *HVAC value meal* quotes from contractors, so they compare them by price and pick the cheapest option. It's a feedback loop that reinforces the expectations of both contractors and homeowners.

We need to break the cycle. Skilled labour and high-quality equipment matter if we want to use right-sized HVAC as a tool to solve homeowner problems and create feel-good homes.

SCIENCE LESSON #4:
What are BTUs and tons of heating?

How much ice would it take to keep your house cold in the summer? Think about dragging huge chunks of ice into your house to keep it cool on a hot day.

OK, they didn't actually cool houses with ice blocks—but that was the archaic solution for refrigeration. It was called the *ice trade*, where ice was harvested in the winter and stored for summer use.

That's the origin of the measurement that we still use to describe heating and cooling capacity: *tons*. If you have a 1-ton air conditioner, it provides the amount of cooling you'd get by dragging a 1-ton (2,000lb) chunk of ice into your house every day. Heating uses the same metric.[17]

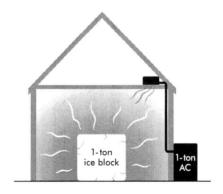

Ice block versus AC

Other measurements—like horsepower and candlepower—are similarly based on real-world observations. They were created at a time when we didn't have a better scientific understanding of the world.

With the creation of mechanical refrigeration, we started using BTU/hr (British thermal units per hour) and kW (kilowatts) as more accurate measurements of heating and cooling capacity. They measure the same thing, just as miles and kilometers are both measurements of distance.

One ton of heating or cooling is equivalent to 12,000 BTU/hr. Therefore, a 3-ton heat pump provides 36,000 BTU/hr or 10.55 kW of heating or cooling.

Red flags from bad contractors

It's easy to find a contractor. They're everywhere, and they're all willing to do your project. The hard part is finding a *good* one.

Here are some red flags from contractors to avoid:

1. Choosing equipment sizing based on a rule of thumb, or skipping the sizing process entirely and using the existing furnace size.

2. They don't measure ductwork dimensions or test duct pressure.

3. They don't discuss problems you're experiencing in the house and how to solve them.

4. They redirect you to gas furnaces with common heat pump myths like *They don't work below freezing*, *They're unreliable*, or *You always need a gas furnace for backup heat*.

5. They try to win the job solely on price. This is a sign that they are in the race to the bottom and are cutting corners with inexperienced labour, cheap equipment, or excluding important add-ons like high-efficiency filtration.

Contractor performance and expertise exists on a spectrum. It's important to choose contractors on the higher end to increase the odds of hitting the tipping point for your house—the point at which the indoor conditions can be confidently controlled. In many cases, installation quality matters more than equipment selection. Even the best equipment performs poorly if it's installed poorly.

Unfortunately, star-rating reviews from homeowners are not reliable indicators of the quality of work. Stars reflect short-term customer satisfaction. It's like choosing a doctor based on bedside manner without considering their long-term success with patients. A friendly doctor who steers you away from vaccines and preventive care is bad for your long-term health, regardless of how much you enjoy chatting with them.

What is wrong with rules of thumb and gas appliances?

I could write an entire book about that.

Isn't this book about that?

Ha—yes, truly. In short: those rule of thumb recommendations are a disservice to homeowners and lead to the wrong solutions. They serve contractors' business models, not homeowners. They're the easy way out. In the field, bad contractors say things like *I've been doing this for twenty years, I know what equipment you need*. My response is, *That's not the brag that you think it is—you've been doing it wrong for twenty years.*

Using a rule of thumb—such as 1 ton of heating for every 500 square feet of living space—is like trying to guess how many words are in a book without opening it. What's the font size? What's the spacing? How many pages are there? Are there a lot of pictures? Bad contractors judge a book by its cover.

It's impossible to know the heat load of a house without measuring it. We have the tools to do accurate heat load calculations, but bad contractors choose not to use them.

What about contractors that recommend gas appliances?

It usually closes the door on the right solution.

Feel-good homes need right-sized HVAC, and for most houses in Canada, right-sized HVAC means heat pumps. The heat loads for most houses in Canada are too small to be well served by gas furnaces—those furnaces aren't offered in small enough sizes for most houses. A gas furnace without a heat pump is almost always the wrong choice.

Hybrid systems (a heat pump combined with a gas furnace for backup heat) are useful options in some houses, but they still need to be the right size.

Getting the calculation right

HVAC sizing comes in intervals of 6,000 BTU/hr (half tons), so "close enough" is the goal for correctly sizing equipment. Energy models have a long list of variables that can be adjusted—the modelled heat load calculation appears *accurate*, but it's actually just *precise*.

Two people try to guess the word count in a closed book: the first picks 30,000 and the second guesses 46,521 words. The second guess is *precise*, but that doesn't mean it's more likely to be correct (*accurate*). The precision of modelled heat load calculations can be misleading for the same reason.

Let's look at an example.

I consulted for a homeowner (ie I was acting on the homeowner's behalf, not operating as an HVAC contractor) in Toronto. It was a hundred-year-old double-brick row house connected to neighbouring houses on both sides. It was quite leaky because of an issue on the top floor.

An energy advisor assessed the house, completed an energy model, and created a full report with recommendations. The report included a heating requirement of 83,052 BTU/hr (6.92 tons) and estimated the house would use 3,971 m^3 of gas per year for heating. The contractor recommended a 7-ton gas furnace based on the results from that energy model.

That's great—it's the right size!

Here's the problem: over the previous twelve months, the house only used 1,300 m^3 of gas for heating—68% less than the modelled amount. I confirmed with the homeowner that they hadn't taken any winter vacations that would have skewed the data. I did a performance-based heat load calculation based on actual gas consumption, and the actual heat load was 26,000 BTU/hr.*

---

* I made the head load calculation based on the gas consumption in January and February, the outdoor temperatures—heating degree days—during that period, and the gas furnace efficiency.

DIFFERENCE BETWEEN GOOD AND BAD CONTRACTORS

The right solution was a 2-ton heat pump, not a 7-ton furnace. The recommended furnace was more than three times too big.

I shared the results with the contractor, and this is what I was told:

- *I have designed hundreds of these systems over the years.*
- *The best guide is common sense and not to take undue risk.*
- *Our business plan is built around providing comfort.*
- *We can certainly assess a heat load in terms of whether it is realistic.*
- *I have designed for many engineers and scientists. Clearly you have to be on the ball to do that.*
- *Having done so many of these and having attended countless training sessions over the years, you get lots of knowledge and experience.*

When I asked a more detailed question about their process to assess heat load calculations

I was told, *I am sorry that I cannot share my trade secrets.*

I'm confident that the homeowner would have chosen the oversized furnace if I hadn't been consulting on the project. That scenario plays out every day. Contractors are trusted experts that are expected to install the right equipment, but they don't.

Nothing would have gone terribly wrong if that homeowner had chosen the 7-ton furnace. In fact, the existing furnace was 80,000 BTU/hr (~6.5 tons), so the homeowner already knows what it's like to own an oversized furnace. The furnace short cycles (turns on and off frequently) to blast hot air through the house, like a bucket of hot water dumped on your head instead of the smooth flow from a shower. For this homeowner, it was causing comfort problems on the top floor because those short bursts didn't give enough time for the system to push air to the farthest rooms.

The homeowner ultimately installed a 2-ton heat pump. When I followed up, the response was simple but powerful: *All is well*.

An ironic thing about right-sized HVAC is that you forget it's there. Homeowners notice uncomfortable rooms and noisy fans, but when those problems disappear, the heat pump works in the background and homeowners don't give it a second thought.

That's the remarkable feeling that feel-good homes create, and it's easy to take that feeling for granted.

Summary

- Homeowners and contractors are entangled in a race to the bottom, where the cheapest quote wins the bid.

- HVAC has become a commoditized product, where high-quality work and remarkable outcomes aren't valued appropriately.

- Most HVAC is replaced after emergency failure in peak seasons (July/August or January/February).

- Emergency replacements are often like-for-like (same model and size as the old equipment), and unsurprisingly, comfort and health problems stay the same or get worse.

- When contractors don't put in the time and effort for accurate sizing, we end up with oversized furnaces.

- Traditional contractors size equipment with rules of thumb and modelled heat loads. Those methods are misleading and tend to be unreliable and inaccurate.

THREE
The Future Of Heating And Cooling

Cold-climate heat pumps are a modern heating solution that can replace archaic options like burning fuel for warmth. With the development of affordable renewable energy, heat pumps provide a path to lower carbon emissions while improving home comfort and air quality. It's a rare win-win decision for homeowners.

In this chapter, we'll discuss the science behind heat pumps, build confidence that they work in cold climates like Canada, and explore how they fit into a sustainable future.

Heat pumps 101: How do they work?

Cold-climate heat pumps can work in outdoor temperatures down to −30 °C (excluding wind chill and humidity). For context, at the time of writing, Toronto hasn't touched −30 °C in more than thirty years (since Sunday 16 January 1994).[18]

For added resilience, heat pumps are commonly installed with backup heating—usually an electric heat strip or gas furnace. Smart thermostats can automatically switch to the backup heat source during extreme cold snaps or heat pump failure.

From my discussions with homeowners, I've realized that most questions about heat pumps come down to one concern: *Will a heat pump be an effective, reliable heat source for me?*

Heat pumps provide remarkable benefits to homeowners because they can be properly sized to load match throughout the year.

Electric vehicles (EVs) are a useful parallel. The accelerating transition from gas cars to EVs isn't led by drivers understanding breakthroughs in battery technology. It's because those breakthroughs enabled EVs to provide superior customer experiences. Heat pump technology had a breakthrough in cold-weather performance that allows for superior homeowner experiences in cold climates like Canada—the ability to create comfortable, healthy, sustainable homes.

There is no way around the fact that fossil fuels are toxic, and switching from a gas furnace to a heat pump also eliminates the risk of gas leaks and carbon monoxide poisoning from your HVAC equipment. Homes are safer and healthier without fuel-burning appliances.

SCIENCE LESSON #5:
But how do you get heat from cold air?

Let's talk about a secret that I don't normally tell homeowners. It's the real answer to the question *How do heat pumps work when it's cold outside?*

Why don't I tell homeowners? Because they don't care. Not really, not in a technical sense. What they're usually trying to say is *Someone told me that heat pumps don't work in Canada. Is that true? Am I going to regret buying one?*

Let's take a deeper dive into the science to get the answers.

Heat is the term we use for thermal energy. *Hot* and *cold* are relative terms that describe differences in the amount of heat: hot means more energy; cold means less energy. When people picture hot and cold, it's based on the small range of temperatures where humans feel comfortable—roughly 20–25 °C. Hot and cold is used to describe any temperature outside of this range.

The temperature scale that we use adds to the misunderstanding. Similar to the origin of *tons of heating*, Celsius (°C) and Fahrenheit (°F) are based on the tools that were available at the time.

100 °F was the best guess for normal body temperature, and 0 °F was the coldest temperature that was easy to reproduce in a lab in the 1700s (the freezing temperature of a saltwater mixture).[19] Celsius is based on water freezing (0 °C) and boiling (100 °C). Those are useful reference points within the historical context, but they don't tell us anything about how much *energy* the different temperatures contain.

So, how do we measure energy? By reintroducing the *Kelvin* temperature scale from high school science class. Kelvin (K) is a measurement of energy and it uses *absolute zero* (0 K) as its reference point. Absolute zero, approximately −273 °C (−460 °F), is the theoretical temperature where particles have zero energy. It's really, really cold.[20]

Homeowners assume that cold air doesn't have much energy, but is that true? Let's compare warm summer air (27 °C/80 °F) with cold winter air (−13 °C/8 °F). How much more energy is in warm air? Is it double? Triple? Ten times?

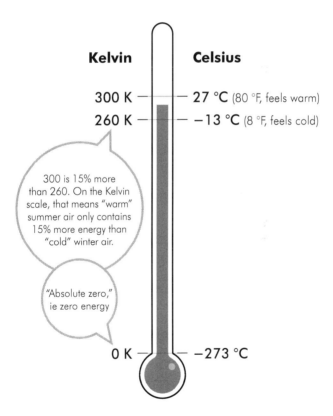

The Celsius and Kelvin scales

We can compare energy levels by converting those temperatures to Kelvin: 27 °C is 300 K and −13 °C is 260 K. The Kelvin temperatures show that warm air only contains about 15% more energy than cold air (300 / 260 = 115%). That's a surprisingly small difference.

Back to heat pumps: heat (thermal energy) naturally flows from hot to cold. The outdoor part of the heat pump is a refrigerant coil that can reach temperatures of −35 °C. To get heat from cold air, the heat pump's outdoor coil simply needs to be colder than the surrounding air. Even during a frigid −30 °C winter night, heat flows from the air and *warms up* the −35 °C coil. That's how you get heat from cold air.

Fossil fuels aren't future-proof

Burning fossil fuels contributes to climate change, and it's bad for your personal health (air quality) and safety (gas leaks and carbon monoxide poisoning). All forms of combustion carry these risks. In addition to being comfortable and healthy, feel-good homes are sustainable.

Electric heating powered by renewables is the only sustainable way to heat a house. Before you raise concerns about the electricity grid: yes, electricity generation should (and will be) entirely renewable in the future. Advancements in cold-climate heat pumps have made it viable to switch from fuel-burning appliances to fully electric alternatives, a process called *electrification*.

Consider this: you can't generate gas, propane, or oil in your backyard. You can't control the fuel prices that are impacted by global events and international markets. Most houses can, however, generate their own electricity (via solar panels) for heat and power.

Electrification is a path to energy independence.

But what about the grid?

Fine. I'll add a section later to talk about the grid, just for you.

Back to the point: electricity sources *can* be renewable, but fossil fuels *are never* renewable, regardless of the marketing spin around clean gas, hydrogen heating, carbon capture, etc.

The average Canadian house generates 5–10 tons of CO_2e emissions per year. In

Ontario, fuel switching from natural gas furnaces and water heaters to electric options like heat pumps reduces emissions by ~95%.

Does saving 5–10 tons of CO2e matter?

At scale, yes. Emissions from homes in Canada make up 18% of national carbon emissions.[21] That includes off-site generation of electricity to power homes but doesn't account for embedded carbon in the building materials.

Fortunately for homeowners, the path to sustainable homes doesn't require a sacrifice—electrification leads to lower emissions, better homeowner experiences, and energy independence.

Preparing for climate change

There are parallels between our response to the COVID pandemic and to climate change.[22]

COVID is caused by an airborne virus that's invisible to the eye. Its invisibility makes it easy to ignore. Unfortunately, though, an increasing number of people are being disabled by the long-term effects of long COVID, with damage to the brain, lungs, and other organs. Over time, there will be cascading impacts on public health and services, either implicitly or explicitly.[23]

Similarly, we don't see the impacts of climate change in our daily lives. We've reached a tipping point in environmental systems that are leading to more severe storms, longer wildfire seasons, and extreme temperatures.[24]

Extreme weather in your area is a matter of *when*, not *if*. When it happens, will your house be safe, or will it put you in a vulnerable situation, where your strategy becomes *Hope it isn't as bad as the weather channel predicts*?

A feel-good home is more resilient against extreme weather.

Heatwaves are the obvious example, but all building envelope upgrades improve resilience. Heat pumps provide cooling in the summer, providing relief against the growing risk of heatwaves.

More countries are seeing dangerous annual heatwaves with wet-bulb temperatures (a temperature measurement that accounts for humidity levels) above 31.5 °C.[25] At those wet-bulb temperatures, the body loses its ability to cool via sweating, leading to dangerous increases in body temperatures.

I live in a cold climate. Doesn't that mean we'll end up with nicer weather?

I wish it worked that way, but no. Climate change will impact every area differently. The reality is that we are heading toward *extreme weather*, not just warmer weather. Global temperatures increasing on average lead to a chaotic change in weather systems that creates more extreme events.[26] For example, in 2021, the village of Lytton, BC, reached 49.5 °C, breaking a heat record from 1937. It broke that 1937 record for three days in a row.[27]

Is your house prepared for a week-long 40 °C heatwave? It needs to be. Canadians are just as vulnerable as other countries to climate change, and we need to prepare our homes to cope with those events.

Summary

- Cold-climate air-source heat pumps work in Canadian winters.

- In most of the country, no backup heat source is needed. There are edge cases where supplemental or backup heating systems are beneficial.

- Heat pumps provide a path to residential decarbonization by heating without combustion. Homeowners can reduce carbon emissions while improving home comfort and indoor air quality. A rare win-win situation.

- Climate change is creating more frequent severe weather conditions. Most houses aren't prepared for extreme conditions such as extended cold snaps, heatwaves, winter storms, and wildfires. Homes need to become more resilient.

FOUR

The Path To A Feel-Good Home

At this point in the book, you have a decent understanding of the problems that homeowners face. Houses are an imperfect system of systems, but we're ready to take action to use those systems in our favour.

In this chapter we'll discuss the role of HVAC in home performance and learn how right-sized heat pumps can be leveraged as a tool for better outcomes.

The universal benefits of right-sized HVAC

Finding the right HVAC system is a required step to create a feel-good home.

Yeah, I still don't get that—isn't bigger better?

No, bigger is not better. Oversized HVAC equipment is the underlying cause of most homeowner complaints.

The number one benefit of right-sized HVAC is comfort, and there are four main factors for comfort:[28]

1. **Air temperature**—the temperature on the thermostat
2. **Surface temperatures**—the temperatures of the walls, floors, windows, etc
3. **Air movement**—wind, drafts, and fans
4. **Humidity**—the amount of moisture in the air

Having an oversized furnace causes two states: either the furnace is blowing hot air, making homeowners feel uncomfortably hot; or the homeowners turn the furnace off to avoid that heat, which makes the house uncomfortably cold. The furnace can't provide the right amount of heating at any given time.

Right-sized HVAC can be ramped up or down to provide the exact amount of heating or cooling you need. That leads to more consistent air and surface temperatures, and it lets you use a lower fan speed, so you get the right air temperature with slower air movement.

What's the right size for my house?

Every house has an amount of heating and cooling needed to maintain thermostat temperatures. That amount is called the *heat load* or *cooling load*, and it changes every minute, depending on the difference between indoor and outdoor conditions.

On cold winter nights, the house needs a lot of heating. On a spring afternoon with 18 °C outdoor temperatures, there might be no heating or cooling needed. You could open the windows, turn off the HVAC, and be perfectly comfortable for an hour or two until the conditions change.

If the heat load constantly changes, how do you know what size is right?

That's an insightful question. Sizing is based on very *cold* outdoor temperatures, not the absolute worst-case winter scenario, but the 99% worst. That temperature is called the *99% design temperature*. It's the cutoff where 99% of the hours in the year are expected to be *warmer* than that temperature.[29]

Design temperatures vary by local climate and change over time. It wouldn't make sense to size a furnace in Toronto for 18 °C outdoor temperatures (when no heating is needed), or even 0 °C, because winters in Toronto are colder than that. The goal is to get HVAC that's the right size for most of the year.

What about the other 1%?

Everyone has that concern, but 1% of the year is only eighty-seven hours. Most of those hours will be within a few degrees of the design temperature, and many will be overnight when indoor temperatures are naturally cooler. Most houses also have a backup heat source that can provide a boost of heating if you need it.

How cold is the coldest 1%?

It depends on your city. It's −3 °C in Vancouver, −29 °C in Edmonton, −16 °C in Toronto, −21 °C in Ottawa, and −15 °C in Halifax.[30]

OK, so I live in Toronto. How much heat do I need to keep warm at −16 °C?

That question requires a heat load calculation that uses consumption data or thermostat runtime data to calculate how your house performs in real conditions.

Any heat load calculation that's strictly theoretical, with a rule of thumb or energy model, is ultimately a guess. If the assumptions are accurate, then it might be a good guess. If the assumptions are wrong, then it's a bad guess—in data analytics, that's called *garbage in, garbage out*.

SCIENCE LESSON #6:
Heat flow

When you sit beside a window on a winter day, you can feel cold even if the thermostat says 23 °C. Why are you uncomfortable?

Every surface radiates heat, but cold surfaces radiate a small amount and hot surfaces radiate more. It feels like the window is *pulling* heat from you, but you're actually *radiating* (pushing) heat to it.

It acts like water on each side of a dam. The water (heat) wants to flow from the high side (hot) to the low side (cold). A dam is effective at stopping water, but heat is more difficult to control. That's the primary role of insulation and air sealing—to slow the transfer of heat.

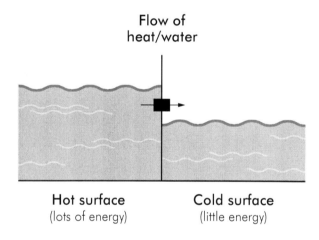

Flow of heat/water

If surface temperatures are close to the temperature of the air, there's less heat flow, which leads to better comfort.

The average temperature of the surfaces around you is called the *mean radiant temperature*, and it's one of the main comfort factors in your home.[31]

Why right-sized means heat pumps

As I outlined above, heat loads change by the minute. When the temperature drops overnight, more heating is required to keep indoor temperatures constant. The same applies to summer temperatures—more cooling is required in the early afternoons when it's hottest outside or the house is sitting in full sun.

We need an HVAC system that can match the changing heat and cooling loads throughout the year. This is called *load matching*.

Heating and cooling loads depend on the local climate and specific house, but most homes in Canada require 2- to 3-ton HVAC systems. Residential heat pumps range from 1.5 to 5 tons (18,000–60,000 BTU/hr), and the most common gas furnaces are 5 to 10 tons (60,000–120,000 BTU/hr).

So, why heat pumps?

They don't make gas furnaces small enough to be the right size for an average house in Canada. If your house has a 2-ton heat load, your HVAC choice is between a 2-ton heat pump or a 5-ton gas furnace. One of those is right-sized, and the other is oversized.

For that reason, right-sized HVAC for the average Canadian house needs to be a heat pump.

Can I get a heat pump and a furnace?

Yes. That's called a *hybrid* or *dual-fuel* system. It's useful with high heat loads or when you're in an extremely cold climate like Edmonton.

In Toronto, the recommended backup option is a resistance heat strip. It's similar to baseboard heating, but safer and more convenient. An electric heat strip is installed in the air handler, above the fan, and lets the system act like an electric furnace during extreme cold snaps or heat pump failure.

The other important considerations for sizing heat pumps properly are ductwork and electrical panel capacity. All of these are included in the HAVEN method.

Less (maintenance) is more (value)

There's a gap between the amount of maintenance that homeowners *should do* and *what actually gets done*.

How do you know?

Maintenance gets ignored until the equipment breaks, and then the homeowner calls a contractor to fix it. I'm frequently on the other end of that phone call.

I don't blame people for ignoring nonurgent issues. Homeowners have busy lives, limited time, and strict budgets. However, we need to start choosing solutions that align with that reality. As often as possible, homeowners should opt for low- or no-maintenance options.

For HVAC, that means getting one piece of equipment to improve comfort and air quality. Adding that filtration and ventilation to a right-sized heat pump lets it solve problems without increasing maintenance requirements. It's one unit to buy and replace, and only one filter to change.

I can't remember the last time I changed a filter.

Yeah, that's my point. That's the bare minimum maintenance that should be done. Gas furnaces are a health and safety risk if the heat exchanger isn't properly maintained. Heat pumps are fundamentally safer because there's no combustion, so there's no risk of gas leaks or carbon monoxide poisoning.

Summary

- A heat load calculation tells you how much heating is required (ie the size of heat pump you need) to keep a house warm on a cold winter night.

- Most houses in Canada have heat loads between 2 and 4 tons (see Science Lesson #4 to learn about *tons of heating*).

- Gas furnaces are generally 5 tons or more, which makes heat pumps (1 to 5 tons) the only option for right-sized HVAC in average Canadian houses.

- Homeowners avoid or delay home maintenance, so it's better to install right-sized heat pumps to provide heating, cooling, dehumidification, filtration, and ventilation in one package to minimize maintenance requirements.

PART TWO
FINDING THE RIGHT PATH

FIVE
Introduction To The HAVEN Method

You're on your way to creating a feel-good home that's comfortable, healthy, and sustainable. Just by reading this far, you know more than 99% of homeowners. The HAVEN method, created as a path to a better home, will help with the next steps.

Do you want good news or bad news first?

Ugh. The bad news.

The bad news is that your house isn't perfect—it's some combination of uncomfortable, unhealthy, and unsustainable. The good news is that we can fix it.

The steps to creating a HAVEN

The challenge is that all houses and homeowners are different. There are different house types, shapes, sizes, climates, underlying problems, homeowner goals, and budgets. The quality of a house depends on the original builder, long-term maintenance, and every renovation along the way. Some buildings fail after twenty years while others last for centuries.

Our understanding of building science has also changed dramatically over time, and building codes have slowly followed. A 1940s house might be a patchwork of outdated band-aid solutions, and the HAVEN method offers modern materials, techniques, and a better scientific understanding to solve problems that have built up over decades.

So, what do we do?

HAVEN was created to simplify the process for homeowners.

HAVEN stands for:

- **H**—heat load
- **A**—air leakage

- **V**—value mindset
- **E**—environmental control
- **N**—necessary infrastructure

HAVEN is spread over four steps:

1. **Assess** the house to understand its needs
2. **Shift** your mindset to focus on value and outcomes
3. **Plan** for right-sized HVAC and other upgrades
4. **Implement**—*do the work!*

Before we walk through the process in the following chapters, there's an exit ramp for some homeowners. It's an even easier path—a shortcut of sorts.

The shortcut for simple homes

Houses can be categorized into three groups: *heat pump-ready*, *nearly ready*, and *needs work*.

Simple homes are *heat pump-ready*. "Simple home," in this context, is a compliment—the problems and solutions are straightforward.

A simple home can install a right-sized heat pump to solve its problems without other upgrades.

The other end of the spectrum is a house that *needs work*. It could require an invasive *deep energy retrofit*, like substantial air sealing and insulation upgrades, to solve its problems. This type of house tends to be a good option for a hybrid system (a right-sized heat pump with a backup furnace) to improve homeowner problems without fully solving them.

A house that's *nearly ready* is somewhere in between.

The biggest difference between house types is the heat load. *Heat pump-ready* homes tend to need 2.5-ton heat pumps (or smaller) and they have simpler problems to solve. Check out www.foundryheatpumps.ca/resources for an assessment to see if your house is a good fit for the shortcut.

Summary

- The HAVEN method includes five categories designed to guide the conversation between contractors and homeowners: heat load, air leakage, value mindset, environmental control, and necessary infrastructure.

- Homeowners and contractors need to assess the house, find the right mindset, create a plan, and implement projects that are likely to solve underlying problems.

- There are three types of homes: *heat pump-ready*, *nearly ready*, and *needs work*. Simple homes are *heat pump-ready*, and they're easiest to transform into feel-good homes. The other types often require more complicated projects.

SIX
Step 1: Assess The House

The first step is a problem-oriented assessment. It focuses on identifying and understanding the needs of the house. In this step, we don't consider the homeowner's goals or budget. We only care about how the house performs to find out what it needs. It is an objective health checkup for your house—let's run some tests.

As you could have guessed, we need an accurate heat load (H) to understand how the house performs during the coldest days of the year. A performance-based heat load calculation informs what *right-sized HVAC* means for the house. This book focuses on cold climates, but this step could include a similar calculation to find the *cooling load* for houses in warmer climates.

Houses with large heat loads or significant problems will look next at their air leakage (A). Air leakage is tested with a *blower door test*, which uses a fan and sensors to measure leakage at different air pressures. Air leakage is the biggest source of heat loss, and leaky houses perform exponentially worse at extreme temperatures than tight houses.

Houses go through a basic assessment when they are bought and sold—while it's good that home inspections cover major issues like structural risks, water damage, and electrical or plumbing issues, the inspectors never talk about how the home will *perform* for its new occupants. Real estate listings should include actual energy use, heat load calculations, and air leakage test results instead of "walkability" scores, room counts, and flowery sales language.

The reality is that most houses have never had a performance assessment, at least not in a meaningful way. This chapter explains how you can use heat loads (H) and air leakage (A) as a starting point to understand what your house needs.

H—Heat load

This book has emphasized the importance of right-sized HVAC. An accurate heat load calculation is the only way to know the right size. Heat load is the amount of heating *your* house needs at a specific outdoor temperature. Its purpose is to calculate how much heating is required to keep a house comfortable during the coldest winter nights.

To recap, the outdoor temperature on that *cold winter night*, which is used for the heat load calculation, is called the *design temperature*, and it's common to use the 99% coldest temperature of the year. Only 1% of the hours are colder. In Toronto, at the time of writing, the 99% design temperature is −16 °C.

There are three types of heat load calculations: rules of thumb (educated guesses), energy models (theoretical), and performance-based (real-world data).

As we've discussed, most contractors use rules of thumb to decide on HVAC solutions, but this is a lazy shortcut that leads to bad results.[32]

The most common rule of thumb is based on square footage (sqft). Contractors estimate the size of the house and do quick math like *1 ton of heating per 500 sqft*. They'll put a 5-ton furnace into a 2,500 sqft house and 8 tons into a 4,000 sqft house, regardless of whether the house is leaky or tight, uninsulated or insulated, two years old or twenty—the recommendation is based on square footage.

The potentially worse (and more common) practice is to skip the sizing process altogether. Many traditional HVAC companies will simply replace an *old* 8-ton gas furnace with a *new* 8-ton gas furnace without questioning the size.

What's the right size?

There's no easy answer to that—it depends on the house. Trying to guess the right size is like guessing a car's mileage without looking at the odometer on the dashboard. If you know the age of the car, you could guess with a rule of thumb, for example 10,000 km driven per year. You might get lucky, but it's still a guess.

What about energy modelling?

STEP 1: ASSESS THE HOUSE

Energy models are data-driven software simulations. They are used by engineers, energy advisors, and HVAC designers to predict energy use. They're based on real measurements, so it's a step in the right direction. Some energy models are quite complex, but the high number of inputs is only valuable if data collection is accurate.

The most common types of modelling in Canada are F280, Manual J, and HOT2000 (H2K). The latter is used for NRCan energy assessments under the EnerGuide Rating System (ERS).

Energy models calculate heat load based on inputs like wall and roof dimensions, construction style, quantity and orientation of windows, and insulation values. Many models don't include measured air leakage, though, which—as we've discussed—is the biggest factor for heat loss. I've seen modelled heat loads that are three times higher than the actual heat load.

Models give the illusion of accuracy, but there is no way to know if the output is accurate without comparing the results to real-world performance.

What was the third option?

Performance-based heat loads. It's the only method that's accurate and reliable. Energy models are greatly improved by adjusting the results to align with actual performance.

There are two types of performance-based heat load calculations: energy consumption and runtime data.

Method 1: Energy consumption

Energy consumption (also called *energy usage* or *gas usage*) looks at how much gas (or other fuel) is used to heat the house. Unlike rules of thumb and energy models, energy consumption is based on house performance in real-world conditions.

In the car mileage example, it's like using gas station receipts (fuel consumption) and the car's fuel efficiency to calculate the mileage.

For a house, you can take gas consumption from a winter month (January and February) and compare it to outdoor temperatures over that period to see how much heating was required. The relationship between gas

STEP 1: ASSESS THE HOUSE

consumption and temperature differences (*heating degree days*) tells you the *incremental heat load*—how much heating was needed to warm the house by one degree Fahrenheit. The *incremental heat load* and the *design temperature* are enough to say how much heating the house needs on a cold winter day.[33]

Method 2: Runtime data

Runtime data shows how long a furnace operates over a period of time. Some equipment and thermostats give access to runtime data. It's best if you can get runtime during an hour when outdoor temperatures are close to the design temperature.

By definition, right-sized HVAC should run close to 100% capacity at design temperatures. If you live in Toronto and it's −16 °C for 60 minutes, the furnace should be operating at full capacity for all sixty minutes. That's right-sized. If a furnace runs for only thirty of those sixty minutes, it only used 50% of its capacity during one of the coldest hours of the year. A furnace that uses 50% capacity is two times too big. I've come across runtime data where the furnace runs for only fifteen minutes at design temperature—that's four times oversized.

Why is that bad?

If the furnace is four times oversized during the coldest hours of the year, imagine how much worse it is during the rest of the year. The daily average on a winter day in Toronto is −3 °C. If the furnace runs for fifteen minutes per hour at −16 °C, it probably runs for five minutes or less at −3 °C. The situation is worse during shoulder seasons, when houses only require a small amount of heating.

Of the two performance-based options, my preference is energy consumption because it's easier to get a monthly gas bill than thermostat data. Runtime data can also be difficult to interpret for multiple-stage or variable furnaces.

Do I have to do my own heat load calculation?

No, you shouldn't have to. You *can*, and I know homeowners that *did*. My best advice, though, is to find a contractor that understands the importance of accurate heat loads. If you find a contractor that values right-sized HVAC, it's a good sign that they'll understand and implement other important details.

STEP 1: ASSESS THE HOUSE

There is a reason that HAVEN starts with *heat load*: you can't have right-sized HVAC without knowing the heat load, and you can't have comfort without right-sized HVAC. An oversized heat pump, furnace, and AC won't cut it—they remove you from the path to a feel-good home.

If you only take one thing from this book, it's the importance of performance-based heat load calculations for positive homeowner experiences.

A—Air leakage

The only way to know the air leakage of a house is to measure it. Air leakage is an important variable for houses with high loads or serious problems. There are two benchmarks for air sealing: leakage-to-air infiltration ratio (LAIR) and air changes per hour (ACH).

Method 1: Leakage-to-air infiltration ratio (LAIR)

Nate Adams—an expert in building science and HVAC, and the writer of the foreword for this book—developed a metric called LAIR (leakage-to-air infiltration ratio). It compares measured air leakage to square footage. After measuring air leakage and tracking energy performance in countless homes, Nate noticed the tipping point was around a one-to-one ratio.[34]

That is the point at which it is usually possible to control indoor conditions in a house without building envelope upgrades.

A one-to-one ratio means the air leakage results from a blower door test, measured in cubic feet per minute (CFM), is equal to the above-ground living space (in square feet).

For example, a house that's 1,500 sqft (excluding any basement) with 1,500 CFM blower door measurement has a one-to-one LAIR and can likely be improved with right-sized HVAC.

Most heritage and century homes have blower door numbers that are one and a half to two times higher than square footage. I've learned from experience, though, that you can't guess air leakage by the age of the house. It needs to be measured.

STEP 1: ASSESS THE HOUSE

Method 2: Air changes per hour (ACH)

Joe Lstiburek—another expert in the building science industry—says that "if you get below 3.0 [air changes per hour], the comfort problems go away, things become predictable, and you save energy."[35]

Air changes per hour (ACH) is based on the same blower door test. It compares air leakage rate with the *volume* of the house. Joe's recommended sweet spot for air sealing is 3.0 ACH, which is achievable in leaky homes by focusing on the biggest holes in the building envelope. It can take an invasive project to find and seal those holes, though.

Nate and Joe both recommend testing the house using a blower door test to collect performance data and using the data to make informed choices. LAIR versus ACH is a matter of preference—do you like apples or oranges?

Is one of them better?

I like apples more.

No, I mean for air leakage.

Oh, no, one isn't better than the other. Nate and Joe came to their answers independently with field observations. They're both useful thresholds, but our ultimate goal isn't to hit an air tightness score—we just need air tightness to be *good enough* to effectively control indoor conditions and create a feel-good home.

LAIR and ACH provide context for our blower door test results. Find a great contractor that owns blower door equipment and see how your results compare to 3.0 ACH and one-to-one LAIR.

Houses that hit the recommended thresholds from Nate or Joe are usually good enough to solve problems by installing a right-sized heat pump without other energy upgrades. Houses with low to medium heat loads and no serious complaints can skip the air tightness test altogether.

What about houses that aren't close to those numbers?

The higher the air leakage numbers, the more likely it is that the house falls into the *needs work* category. Homeowners typically know if their house is leaky. It leads to discomfort, humidity issues, high energy costs, and high heat loads.

Air sealing can be paired with other projects such as attic insulation, siding replacement, and kitchen or bathroom renovations. Over time, right-sized HVAC and a few strategic renovations could be enough to hit the tipping point.

Summary

- Performance-based heat load calculations use actual data to assess how a house performs under real-world conditions. The two methods are *energy consumption* and *runtime data*.

- Without an accurate heat load calculation, contractors can't know what size of HVAC equipment is the right size. Traditional contractors tend to use rules of thumb or energy models to guess sizing, although many skip the sizing process altogether.

- Blower door tests measure air leakage in a house. Air leakage is the biggest source of heat loss, so it's important to understand, especially for houses in the *needs work* category. For houses with low to medium heat loads, this step is optional.

SEVEN
Step 2: The Right Mindset

Traditional HVAC business models fall short and miss the opportunity to use right-sized HVAC as a tool to solve problems that matter to homeowners. All too often we focus on quick and easy solutions, but these lead us down the wrong path, resulting in the like-for-like replacement of basic, oversized equipment.

To find the right projects and achieve remarkable results, homeowners and contractors need to focus on value, not price.

V—Value mindset

The success of every project depends on the *value* it provides, not the price that it costs. Finding the right mindset is the step in the process between assessing the house and planning the right upgrades.

What is the difference between price and value?

Price is an *input*. Price includes how much the project costs and the monthly payments involved.

Value is an *output*. It's the problems solved, the remarkable results, and the lasting benefits achieved.

Price versus value leads us to different projects. If a homeowner focuses on price, they'll pick the cheaper quote—it usually doesn't matter what is included in each option. To win business strictly on price, contractors are incentivized to pitch cheaper equipment, use cheaper labour, offer shorter warranties, and take shortcuts during the installation process. If you pick contractors based on the lowest input (price), don't expect the highest output (remarkable results).

Value-focused customers complete projects and pick contractors based on their ability to solve specific problems in the home.

Value-based solutions lead to feel-good homes

If you suffer from seasonal allergies, how much is it worth to breathe fresh air without a runny nose, itchy eyes, and constant sneezing? The price-focused solution is a generic allergy medicine. The value-focused solution is upgrading your HVAC system to fix the IAQ problem—investing in clean air with better filtration. It could significantly improve your comfort and quality of life at home without medication, but that project only happens if you're focused on the value of living in a feel-good home.

The ABCs of home upgrades

The success of any home upgrade depends on aim, budget, and complexity:

- **A—Aim:** What is the purpose or goal? What problem are you trying to solve? What outcome are you trying to achieve?

The answers to these questions deliver the target outcome—the aim.

- **B—Budget:** What is it worth to solve the problem? How much are you able and willing to spend? Your answers give the available input—the budget.

- **C—Complexity:** How complicated is the problem we're trying to solve? How likely is the next project to get us to the tipping point? The answer determines the complexity of the problem and solutions involved.

While the aim and budget are based on the *homeowner*, complexity is based on the *house*. Finding a viable project that's likely to solve the underlying problem depends on these ABCs.

It's easy to see how aim and budget need to align—a project doesn't get completed if the outcome isn't worth it or the budget isn't available. Complexity is equally important, though.

For example, let's pretend that you find black mould on the baseboards in your bedroom. You could wipe it away with vinegar. That

STEP 2: THE RIGHT MINDSET

appears to solve the problem, or at least makes it easier to ignore. Alternatively, you could hire a mould remediation company to open the walls to remove all traces of the mould.

Mould remediation is a more expensive and invasive project, but is it likely to solve the problem in the long term? Probably not. The underlying problem is that parts of the house have the ideal conditions for mould to grow.

Removing the existing mould doesn't change those ideal conditions. Is humidity too high in the winter? Is air leakage causing condensation on a cold surface? Is the existing mechanical equipment helping or hurting the conditions? The full solution might require multiple contractors, long timelines, and disruptive work.

There aren't silver bullets for home improvement—a house is a complex system of systems, and solutions can be equally complicated. If you're lucky, an oversized furnace is the only thing stopping you from living in a feel-good home. In many cases, though, there's more to it.

The HAVEN method requires homeowners and contractors to stay focused on the target outcome—what problems are we trying to solve, and what is the best path to get there?

What does that mean for me? Should I still get a heat pump?

Right-sized heat pumps solve a lot of problems. Whether right-sized HVAC by itself is likely to reach that tipping point depends on the ABCs—it might be one of multiple steps on the path.

Every house has a tipping point where it shifts from uncontrollable to controllable. How much is it worth to live in a house with remarkable comfort, clean air, and sustainability? Focus on the *value* of the projects that bring you closer to that outcome.

Summary

- Price is an *input*. Price includes how much the project costs and the monthly payments involved.

- Value is an *output*. It's the problems solved, the remarkable results, and the lasting benefits achieved.

- Prioritize outputs (value) over inputs (price) to get the best results.
- The ABCs (aim, budget, and complexity) help homeowners to find viable projects that are more likely to solve their problems.

EIGHT
Step 3: Create A Plan

Planning is solution-oriented. A feel-good home requires the ability to control indoor conditions like thermal comfort and air quality.

This chapter discusses how to gain environmental control (E) with the right equipment, and we'll look at challenges around the necessary infrastructure (N) to fit those solutions into the house.

E—Environmental control

Indoor environmental quality (IEQ) is fundamental to feel-good homes. IEQ includes air quality, thermal comfort, and other factors like acoustics and lighting that impact your experience indoors.

For comfort, we need to control air temperature, surface temperatures, air movement, and humidity levels. Once air sealing in your house is *good enough*, then you can likely gain environmental control by installing a right-sized heat pump.

Load matching reduces temperature spikes, stratification (temperature differences between floors), and promotes consistent air and surface temperatures, while the longer runtimes result in better dehumidification performance.

Right-sized HVAC is a priority upgrade in every house.

But does it help with air quality?

It can, absolutely.

Traditional HVAC companies sell standalone equipment for each problem. They solve dehumidification with a whole-house dehumidifier; ventilation with an energy recovery ventilator (ERV); and filtration with a standalone HEPA

STEP 3: CREATE A PLAN

filter, UV purifier, or a different type of electronic air cleaner.

That stretches the budget across multiple pieces of equipment. Instead of one heat pump, you end up with five systems (furnace, AC, dehumidifier, ventilator, and HEPA filter). That adds more fans, filters, and failure points. Eventually, it leads to five equipment failures, five service calls, and the cost of five replacement units.

There's a time and place for standalone options. In most cases, though, the job can be done by a single unit with minimal maintenance. With a few upgrades, a right-sized heat pump can do the work of all five.

- **Filtration:** upgrade the return air drop, and install a thick (at least 4"), pleated MERV 13 (or better) filter. Test duct pressure before and after it gets installed to ensure good system performance.

- **Ventilation:** install a fresh air duct to allow your heat pump to distribute filtered outdoor air throughout the house. Add a ventilation controller to make sure you're getting fresh air at the right time.

- **Dehumidification:** choose a heat pump with performance specs that align with the humidity needs of your house or pick a unit with reheat dehumidification. Completing an accurate heat load calculation and installing a right-sized heat pump will naturally improve dehumidification performance.

A right-sized heat pump with a high-efficiency filter and fresh air duct is the perfect foundation for great indoor air quality. That's one piece of equipment. One filter to change. One installation, one service call, and one replacement.

That's the ideal solution for houses that are *heat pump-ready* or *nearly ready*. For any house that *needs work*, other energy upgrades are required before a heat pump can help solve underlying problems. The setup also works well in larger houses if a gas furnace is used for backup heat—this is a step in the right direction for homeowners that aren't prepared to do a *deep energy retrofit*.

STEP 3: CREATE A PLAN

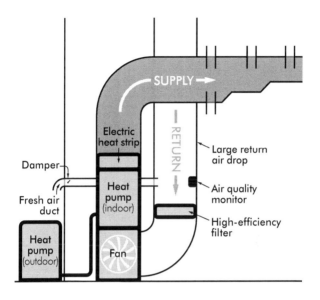

All-in-one HVAC

And you mentioned IAQ monitors before. Do I need one?

IAQ monitors are optional but useful. You can buy consumer-grade monitors and install them in rooms throughout the house to compare pollutant levels. My recommendation is to install a central monitor in the return air duct near your right-sized heat pump to monitor the air as it passes through the system.

My preference is central monitors for several reasons:

- They give a centralized, whole-house view of IAQ.
- They're less intrusive.
- They provide valuable insights into HVAC performance.
- They can alert the contractor when the system isn't operating as intended.

A central IAQ monitor can provide useful information about indoor conditions and HVAC performance with little effort required by the homeowner.

What about houses without ducts?

Ductless heat pumps are amazing at improving comfort throughout a house, but they are severely limited in their ability to improve IAQ and control humidity levels. Most houses with baseboards or radiators have no filtration, ventilation, or dehumidification systems.

The best option in this scenario is usually a *ventilating dehumidifier* with a MERV 13 filter.

It won't last for as long as a heat pump, but it'll perform the IAQ tasks well.

For best results, it should be ducted to each floor.

N—Necessary infrastructure

The major components in a house are the infrastructure that lets it operate as a system of systems. That includes the foundation, walls, roof, and insulation, and mechanical systems like plumbing, electrical, and HVAC.

The final piece of the puzzle is understanding how the recommended equipment fits within existing infrastructure.

When we talk about installing a right-sized heat pump, the two main limiting pieces of infrastructure are ductwork and the electrical panel. Houses without ductwork don't need to worry about that factor—but that's a tradeoff for greater concerns about IAQ and humidity control.

A central ducted system uses a furnace or air handler (located in the basement or a utility room) to push hot or cold air through a network of connected ducts. If you have floor vents, you have ductwork.

I have ducts. Is that good or bad?

We're not worried about good versus bad ducts, although certainly there are leaky, poorly designed, and disconnected ducts. Those are bad. Otherwise, our concern is the *quantity* of ductwork, not necessarily the *quality*.

Ductwork is like a series of straws running from the central fan to each vent. Larger straws can move more air than smaller straws. The furnace or heat pump fan is designed to push a set amount of air, so if you don't have enough ductwork, it's like trying to move a bowling ball by blowing through a straw.

Moving a lot of air through small straws increases the duct pressure; and high pressures lead to poor performance, noisy systems, and early equipment failure.

Why haven't I heard about duct pressure before?

That's because ductwork isn't a limiting factor for gas furnaces. Gas furnaces work at higher temperatures, so they don't need to move as much air.

Imagine if you had to keep a bathtub at a constant water temperature. You could either add a small amount of very hot water a few times each hour, or a slow but stream of warm water. The same amount of energy is added in each case.

The distinction between gas furnaces and heat pumps is similar—your house is the bathtub. Gas furnaces push a small amount of very hot air (50 °C) until the thermostat is satisfied. Heat pumps push a steady stream of warm air (35 °C) to keep the house at a constant temperature. Because gas furnaces don't need to move as much air, they don't need as much ductwork.

In the bath example, you're choosing between a trickle of warm water that keeps the bath at a constant temperature and continues to mix the water versus turning the tap on and off to add much hotter water to one end of the bath while the other end stays uncomfortably cold. With a gas furnace, the proverbial "other end of the bath" is the room that's farthest from the furnace, and that's why it stays cold all winter.

Here are two important truths about HVAC and ductwork:

1. Heat pumps need more airflow than gas furnaces to move the same amount of heat.
2. Ductwork can handle a maximum volume of air before there are problems with high duct pressure.

At the beginning of the HAVEN method, we calculated the amount of heating we need—the heat load—and now we need to check whether there's enough ductwork in the house to move that much heat.

What if there's not enough ductwork?

The house might need a hybrid system to get the steady stream of warm air from a heat pump with the higher temperatures of a gas furnace.

If a performance-based heat load calculation tells us that a house needs 4 tons of heating

STEP 3: CREATE A PLAN

at design temperature, but the ductwork can only handle a 2-ton heat pump, then the easy option is to install a 2-ton heat pump with the smallest available gas furnace (probably 5 tons). The heat pump will carry most of the heat load, and the higher capacity of the gas furnace will cover the coldest days.

If the gap was smaller (for example, 2 tons of ductwork with a 2.5-ton heat load), then a 2-ton heat pump with a supplemental heat strip is the better option. A heat strip can run at the same time as the heat pump, adding a small boost of supplementary heat when it's needed.

The alternative, which might be required if there are other problems in the house, is to plan other energy upgrades to reduce the heat load until it aligns with the capacity of the ductwork.

A traditional HVAC company would recommend a different path, of course. A house with the 4-ton heat load probably has an existing 8-ton gas furnace in it. The recommendation would be to install a new 8-ton furnace. That option is fine from a ductwork perspective, but it misses the benefits of right-sized HVAC.

In every house with a central ducted system, the capacity of the ductwork needs to be considered as a limiting factor. As often as possible, we need to work within the existing infrastructure of the house—it's expensive to replace or upgrade infrastructure.

Does a heat pump need a new electrical panel?

In general, no, it doesn't require a panel upgrade. This depends, though, on what you're installing, what appliances are in the house, and the size of the existing panel.

Panel capacity is measured in amps, and most houses have 100-amp panels. For a "normal" house with an electric water heater, induction cooking, an electric car, and a dryer, you'll be fine to install a 2.5-ton heat pump and 5 kW heat strip.

Houses with atypical loads—like hot tubs, or multifamily homes with several of each appliance—will need to upgrade to 200-amp service, though.

Foregoing a backup heat strip, or choosing energy-efficient appliances like heat pump water heaters and ventless dryers can help to alleviate panel constraints.

Summary

- Right-sized equipment is required for good thermal comfort.

- High-efficiency filters and fresh air ducts can be added to create an all-in-one system that performs all the functions required for good IAQ.

- One high-quality, right-sized heat pump is better than five separate pieces of mechanical equipment.

- Existing ductwork is a limiting factor when sizing equipment.

- Ductwork can move a limited amount of air. Installing a system that's too big for the ductwork will result in high duct pressure, poor performance, noise, and early equipment failure.

- Electrical panel capacity is another limiting factor to be considered.

- Most houses can install a heat pump and electric water heater without changing the 100-amp electrical panel, although special cases like multifamily homes or atypical loads like hot tubs may force an upgrade.

NINE

Step 4: Implement— Do The Work

Creating a plan puts the homeowner on the right path. Remarkable results don't come from a good plan, though. Results come from actions—from *doing the work*. This chapter details what executing a plan looks like, and tips to enjoy the journey.

Execute the plan

Once you've worked through the HAVEN method, you'll know the right heat pump for your house, and whether a heat pump by itself is likely to solve your problems.

Your plan will be one of three types:

- Simple (install a right-sized heat pump)
- Medium (heat pump with complications, electrical work, or a series of other small projects)
- Complicated (heat pump with significant building envelope improvements—a *deep energy retrofit*)

When HAVEN is followed, any of these paths leads to a feel-good home because you designed the projects with outcomes in mind. Now it's time to do the work.

Embrace each step

You don't need to do it all at once, but it's important to start. If you want to start small, it's OK to use HAVEN as an iterative loop, where you assess, plan, and implement a smaller project to see if you hit a tipping point in the house. If not, then it's time to reassess, create a revised plan, and take the next step—continue that process until you've created a feel-good home.

There isn't always a right or wrong order to complete upgrades, but it's important to understand how each project will fit together. If a house has a 3-ton heat load, don't install a 3-ton heat pump right now if you're planning to do air sealing and insulation upgrades next year—your right-sized heat pump will be oversized when the projects are completed. It's fine to install the heat pump first, but make sure it fits within future plans so you don't need to undo or redo any upgrades.

Remember to embrace each step, big and small, as a valuable part of your journey toward a feel-good home. Celebrate your progress along the way.

My house, before and after

HAVEN can be applied to any house, but I created it to make my life better—to make it easier to transform my house into a feel-good home.

My house in Port Hope, Ontario, Canada was built in 1920. It is 1,750 square feet (sqft) of living space—1,300 sqft on the two main floors and 450 sqft in a converted attic, plus 650 sqft of unfinished basement, below grade.

It's a double-brick house, which means the walls are built with two layers of brick that support the rest of the house. It's solid masonry construction with no insulation in the walls. The uninsulated basement had concrete walls, and the attic was leaky and under-insulated. It had original single-pane windows with wavy glass—charming but inefficient—and the air leakage was 6.9 ACH (air changes per hour).

The house was hot in the summer and cold in the winter. It was drafty. The attic (450 sqft) was only usable during shoulder seasons. We closed the attic door for the other nine months and used it for storage. There was a 100-sqft room on the second floor with similar comfort problems—it didn't have ductwork for heating or cooling, so it went unused for most of the year. That's 550 sqft of unusable space—nearly a third of the house (550 of 1,750 sqft).

For mechanical equipment, the house had a 7-ton gas furnace, gas water heater, gas oven and stove, open wood-burning fireplace without a damper, and no central AC. There was no ventilation and the filter was a traditional low-efficiency 1-inch filter—the type of filter designed to catch dust, not improve IAQ.

STEP 4: IMPLEMENT—DO THE WORK

With the oversized furnace, the house had severe temperature swings, so it was always too hot (furnace was on) or too cold (furnace was off). The thermostat needed constant attention—it operated like an on–off switch for the furnace.

The air quality was mediocre. The house got stuffy in the winter with the windows closed. It was too leaky to be comfortable, but not leaky enough for "natural ventilation" to benefit IAQ. It had elevated CO_2 levels—a proxy for poor ventilation.

Did you fix it?

Yes. That shouldn't be a surprise. I wouldn't have written this book if my house was still uncomfortable, unlivable in sections, and had poor IAQ with no control over the indoor environment.

What did you do?

I followed the steps in HAVEN to assess the house, find the right mindset, create a plan, and implement it.

H—Heat load

I did a performance-based heat load calculation based on energy consumption. Between September 2021 and August 2022, I used 2,883 m^3 of gas for space heating (85%) and hot water (15%). A negligible amount was used for cooking.

My house used 79-million BTUs of heating that year. Adjusting for the severity of that winter—it was 4% warmer than average—the house had a heat load of 35,000 BTU/hr. It needed about 3 tons of heating at *design temperature* (−16 °C), but it had a 7-ton furnace with 77,000 BTU/hr output. Even in a 100-year-old leaky house, my gas furnace was two times too big.

That was the baseline for my house—I knew that improving air leakage and insulation would reduce the heat load, but it was important to know the worst-case scenario (ie no upgrades).

A—Air leakage

The blower door test result was 6.87 ACH and 2,401 CFM (cubic feet per minute).

With the 3.0 ACH target in mind, the goal was to reduce air leakage by 50–60%. Air sealing near the top and bottom of a house has more impact because of the *stack effect*, so I knew that I needed to target those areas.

V—Value mindset

My goal was always to electrify the house (switch from gas to electric appliances) and remove the gas meter. Through 100 years of homeownership, the house was drafty and uncomfortable with unlivable floors—a third of its living space couldn't be used for most of the year.

The house likely had five or more furnace replacements over that span, and every homeowner chose to replace an old, oversized furnace with a new, oversized furnace. The only significant change was a switch from oil/wood to gas at some point.

The cheapest option in the short-term was always to keep the same type and size of furnace, use window ACs, and abandon rooms and floors for most of the year. None of my upgrades would have happened if I had only cared about price—I focused on the remarkable results that I knew could be achieved with the right projects.

The plan isn't done yet. In the coming years, I'll get a metal roof, install rooftop solar, a bidirectional EV charger, and continue air sealing improvements. I don't need to rush those jobs, though. They're part of the big picture for my house, and they'll be added in due time.

E—Environmental control

The indoor environment wasn't under control. I knew that right-sized HVAC could improve some of the problems, but it would require building envelope upgrades to fully control the house. Four potential upgrades (in addition to a right-sized HVAC) were highlighted: basement insulation, attic insulation, interior storm windows, and additional ductwork.

Once completed, a right-sized heat pump with a high-efficiency filter and fresh duct could be used to solve thermal comfort and air quality problems.

N—Necessary infrastructure

The existing infrastructure was a limiting factor. The old ductwork could only handle a 2.5-ton heat pump, so the other projects were

STEP 4: IMPLEMENT—DO THE WORK

required if I wanted to electrify the house and remove my gas meter.

It was a quantity *and* quality issue. Not only were there not enough ducts, but they didn't move air evenly throughout the house. The ductwork mainly serviced the first floor. A small amount made it to the second floor, and almost no airflow made it to the third floor (attic).

The electrical panel wasn't a concern. The house had a 200-amp panel, and even though the detached garage was converted into an art studio with a large ceramic kiln and its own ductless heat pump, there was no concern about electrical capacity.

Doing the work

The projects and renovations were completed over several years. Insulating the basement provided tremendous value by reducing heat loss through the foundation walls, and the air sealing benefit came from sealing the top of the walls, between the joists.

Closed-cell spray foam was also added to the underside of the roof. This improved insulation levels and significantly reduced air leakage—the spray foam sealed the connection point between the top of the wall and the roof.

I installed interior storm windows on every window in the house to reduce air leakage and improve energy performance, while retaining the house's historic charm.

Did the air sealing work?

Yes. I ran multiple blower door tests throughout the project, and the final result was 2.8 ACH. Success!

I added new ducts to service the uncomfortable rooms and increased supply to the second and third floors, and I cleaned up the ductwork in the basement to improve overall HVAC performance.

Other smaller projects were completed at the same time: fixing bath fans, replacing the gas water heater with an electric tank, replacing the gas stove and oven with induction, and air sealing exterior doors.

STEP 4: IMPLEMENT—DO THE WORK

The right-sized HVAC ended up being a 2.5-ton cold-climate heat pump with backup heat strip. I doubled the return air drop from its original 8 × 18" (2 sqft) to 24 × 24" (4 sqft), installed a high-efficiency 4"-thick MERV 13 filter, tested the duct pressure, and added a fresh air duct for ventilation.

It's nearly perfect. I have a single system with minimal maintenance that effectively controls the indoor environment to make my house the best possible version of itself.

What is the house like now?

It… feels good (pun intended).

I have two and a half floors above grade, with even temperatures, lower energy bills, better air quality, improved resilience, minimal carbon emissions, and only one filter to change.

The biggest value in my case was the additional 25% of living space by taking back the formerly unusable rooms.

It is now a **feel-good home**.

Summary

- Implementation is vital. It's great to assess your house and create a plan, but that means nothing if you don't take the final step and *do the work*.

- I bought a 1920 double-brick house near Toronto, Canada. During its first 100 years, homeowners chose to keep it leaky and uncomfortable. Nearly a third of the living space was unbearably hot or cold depending on the time of year.

- I followed the HAVEN method and transformed the house—with a combination of insulation, interior storm windows, air sealing, and a right-sized heat pump—into a feel-good home.

TEN
"But The Grid!" (And Other Concerns)

Most questions in building science and home performance can be answered by *It depends*. Unfortunately, asking well-meaning friends, family, and colleagues can muddy the waters by introducing myths and misconceptions.

This chapter answers common homeowner questions on the following topics:

- Cold weather, electricity generation, and the grid
- Ground-source heat pumps (GSHPs)

- Replacing other gas appliances (water heaters and gas stoves)
- Building envelope and window upgrades

Each of the subjects deserves more attention, but the conversations require too much depth. I have therefore included only high-level thoughts to point you in the right direction.

Cold weather, electricity generation, and the grid

You keep talking about cold-climate heat pumps, but I live in a really cold climate.

It's common for homeowners to be concerned that *cold climate* doesn't apply to them, and they might be right.

Modern cold-climate heat pumps work down to −30 °C. For context: at the time of writing, Toronto hasn't touched −30 °C outdoor air temperatures in the last thirty years. For the majority of Canadians, heat pumps can confidently provide heating all year without backup or supplement heat. Electric heat strips are common, though, and recommended in areas with design temperatures between −8 °C and −22 °C, including most of Ontario.

The extremely cold parts of Canada can still benefit from right-sized HVAC by pairing a heat pump with a backup furnace. Smart thermostats will automatically switch to the backup heat source when temperatures drop below the minimum operating conditions of the heat pump.

Can the grid handle mass electrification?

Electrification means switching from fossil fuel to electric options, for example, switching gas furnaces to heat pumps or combustion cars to electric cars. More electric appliances mean more strain on the electrical grid. In most cases, the electricity is moved through transmission and distribution lines from the generation source to the house.

Concerns about the grid come mostly from well-meaning homeowners, but it's also a strategic talking point for oil and gas advocates. It's an important question. If the grid can't handle the increased capacity, then electrification as a climate solution is a dead end.

Here are four reasons why grid concerns are overblown:

1. **It isn't a problem right now.** We're talking about a potential scenario where we electrify homes and vehicles faster than we can expand the grid. The important context when considering electrifying your house is that grid concerns are a hypothetical *future* problem based on a worst-case scenario.

2. **We're planning for the increased demand.** Every electricity grid has a central operator. In Ontario, for example, it's the Independent Electricity System Operator (IESO). They have a dedicated team responsible for long-term planning that prioritizes grid reliability. The increased load from residential electrification is part of those long-term plans.

3. **We've expanded the grid before.** We've shown the ability to drastically increase the size of the grid. Ontario roughly *doubled* its electricity generation between 1970 and 1985.[36] Electricity consumption has been relatively flat in the forty years that have followed because population

growth was offset by energy efficiency gains.

4. **We have the tools to do it again.** We have access to the cheapest source of energy in history.[37] The last decade of innovation caused an incredible drop in the costs of renewable energy and battery storage. We can now build new, flexible, affordable electricity generation quicker than ever before.

The next time someone brings up concerns about the grid, remember those four points. The grid is an important topic, but it should have no impact on the decision to electrify your house—it's not currently a problem, we're planning for the increased demand, we've grown the grid before, and we have the best tools in history to do it again.

My grid is dirty—won't electrification increase my emissions?

When we talk about a "dirty" grid, we're talking about carbon intensity—the amount of carbon emissions from electricity generation and transmission. Carbon intensity varies across the country. For example, Alberta is dirtier than Ontario, which is dirtier than Quebec.

Switching from gas, propane, and oil furnaces to heat pumps reduces emissions over the long term as more renewable generation is added to the grid and older technologies like coal- and gas-powered plants are phased out.

We're no longer at the point where the question is *Do we want to pay a premium for renewable energy?* Renewable energy is more affordable than the alternatives. The new question is *Should we pay a premium for coal- and gas-fired power plants?*

Heat pumps provide the only viable path to sustainable heating.

What about nuclear?

Yes, nuclear power is a low greenhouse-gas-emitting resource. Remember, though, when I said that we have access to new, flexible generation cheaply and quickly? I wasn't referring to nuclear power. Building nuclear reactors is risky, slow, and expensive relative to other renewables.[38]

Planning, construction, and commissioning can take a decade or more, and the result is a power plant with limited ability to ramp up and down. On top of that, upfront costs for nuclear power are significantly higher than solar generation. You therefore get longer development times, less flexibility, and centralized generation at a higher cost.

Ground-source heat pumps

Should I get a ground-source heat pump instead?

Ground-source heat pumps (GHSPs) look great on paper. Ground temperatures are more constant throughout the year so GSHPs provide a societal benefit over air-source heat pumps (ASHPs) by reducing load on the grid during peak winter conditions, but they're almost always the wrong choice for homeowners.

In the winter, an ASHP moves heat from the outside air into the house. As the name implies, GSHPs follow the same underlying process but use the *ground* as a source and sink for heat, instead of the outdoor air.

I'll start by dispelling a misconception: ground-source heat pumps are *not* renewable energy. Even though they're frequently called "geothermal," they don't use geothermal energy.[39] GSHPs are no closer to using, creating, or generating renewable energy than your air conditioner or fridge. They simply move heat from one place to another—from inside the house into the ground, or vice versa.

The downside of GSHPs is the requirement to do expensive drilling or excavation to install ground loops. In Ontario GSHPs cost $15,000 to $20,000 more on average than comparable ASHPs. Vertical loops in difficult locations like Toronto are even more expensive.

Aren't they cheaper to operate, though?

Yes, but not enough to justify the upfront costs. That budget is better spent on building envelope upgrades. Over the last five years, I installed basement and attic insulation, interior storm windows, and reduced air leakage by 50% for less than the incremental cost of a GSHP. Those upgrades provide more savings, while also improving comfort and resilience.

GSHPs have useful, niche purposes in district energy (heating an entire neighbourhood with a connected network of GSHPs), commercial projects, and large multiunit residential buildings (MURBs)—but they're the wrong option for homeowners trying to create a feel-good home.

Replacing other gas appliances

Are there other benefits to electrification?

The two other areas affected by electrification are *water heating* and *cooking*.

Like the adage "old homes are drafty," we have low expectations of our water systems. We expect to turn a tap and wait dozens of seconds (or even minutes) before the running water goes from cold to lukewarm.

The most underrated benefit of electrifying a house is the opportunity to reduce hot water wait times by relocating the water heater.[40] At the risk of sounding hyperbolic, reducing wait times from two minutes to ten seconds can be lifechanging.

Most gas water heaters are located next to gas furnaces because water lines are cheaper to run than gas lines. The placement reduced the length of the gas lines, at the expense of plumbing.

When you switch a gas water heater to an electric model, you're no longer tied to the gas line, so it's easier to optimize water heater placement. By moving the water heater closer to the end fixture, you reduce plumbing length and shorten wait times.

Homeowners should also consider drain water heat recovery (DWHR) systems. It is simple copper piping that recovers energy (heat) from warm water as it goes down the drain. It is a double-wall heat exchanger, so the incoming and outgoing streams don't mix.[41]

DWHR systems range from 2 ft to 7 ft in length (longer systems recover more energy), and they replace a section of the existing drain stack. This can be combined with a low-flow showerhead to reduce energy usage by 60–80% during showers. There are no moving parts, so DWHR should last as long as the rest of your plumbing infrastructure, with zero maintenance.

And what about cooking? Aren't gas stoves better?

"BUT THE GRID!" (AND OTHER CONCERNS)

Many media outlets have highlighted the link between gas stoves and the increased risk of childhood asthma—a risk similar to second-hand smoke exposure.[42]

Gas stoves continue to leak methane while they're switched off, and combustion from gas stoves produces carbon monoxide, nitrous oxides, and additional PM2.5, all of which cause respiratory issues.[43] The additional pollutants put extra pressure on the ventilation system.[44]

The misconception that gas stoves are superior is based on a comparison to the old type of electric stoves that use resistance heating through a spiraled element. That's an unfair comparison, though—it's like arguing that horses are faster than cars but using the original Ford Model T as your example. (The top speed of the Ford Model T was 68 km/h,[45] while legendary racehorses like Secretariat reached speeds above 70 km/h.[46])

Modern induction cooking is a completely different technology. It uses magnetic waves to heat the pot or pan directly, leading to a superior cooking experience.[47] Induction hot plates offer a way to try induction cooking without a major purchase.

If you don't like induction, then keep your existing stove but get a high-quality exhaust hood that provides good coverage over all of the burners, and remember to use it every time you cook.

Cooking on *any* stove produces PM2.5 (hazardous, tiny particles) and VOCs (volatile organic compounds—airborne chemicals), so it's important to use exhaust hoods, regardless of stove type.

Building envelope and window upgrades

Do I need to replace windows or upgrade insulation before installing a heat pump?

It's a common misconception that homeowners *need* to upgrade windows, insulation, or air sealing before installing a heat pump—but as we've discussed in the HAVEN method, it depends on the needs of the house. Many houses can install a right-sized heat pump and achieve remarkable results without other upgrades.

Air sealing and insulation projects are valuable, but they're only *required* for houses with high heating loads, ductwork or panel constraints, or complex comfort and IAQ problems.

I hope I've made it clear that air sealing is the top priority, as air leakage is the biggest source of heat loss. While insulation offers diminishing returns, it's impactful in older homes with minimal insulation. With modern building codes, however, adding more insulation is unlikely to solve comfort problems.

High-performance standards like Passive House are the peak of home performance.[48] Passive Houses are feel-good homes, but that level of retrofit is cost-prohibitive—the results are undeniably good, but it's beyond the means of most homeowners. Focus on target outcomes and choose projects that are likely to solve the problems in *your* house.

And windows?

Window contractors have some of the best salespeople in the world, but most windows don't need to be replaced (yet). If you have the budget and you're considering replacing your windows, I suggest you look at other areas in your house to invest that money. It could pay for a right-sized heat pump, air sealing, improving filtration and ventilation, and ductwork enhancements.

Windows are a matter of preference, so pick windows that fit your style and budget. In general, nonoperable windows (ie ones that don't open) are the most energy-efficient, and hinged windows last longer than sliding windows. Where possible, avoid skylights and patio doors because they are prone to leaking (air and water).

Expensive windows that are installed poorly will perform poorly, so focus on finding the best contractor in your area, regardless of which windows you choose.

Conclusion: Your Feel-Good Home

When you completed the self-assessment at the start of Chapter 1, it's likely you recognized that there is significant room for improvement in the comfort and air quality of your house. The silver lining is that solving bigger problems leads to greater value.

In its simplest terms, the HAVEN process is simply about:

- Finding a great local contractor that takes the time to understand your house
- Planning a project that's likely to solve underlying problems
- Implementing the plan—*doing the work!*

What makes a good contractor ... good?

A good HVAC contractor thinks about the sections of the HAVEN method. They might have different names for it, but the fundamentals are the same:

1. **Heat load:** The contractor understands the importance of right-sized HVAC, which requires a performance-based heat load calculation from energy consumption or runtime data to assess how your house performs under real weather conditions.

2. **Air leakage:** If the heat load is high or underlying problems are complex, they do a blower door test to assess air leakage and start considering non-HVAC projects that might be required to improve the house before right-sized HVAC is installed.

3. **Value mindset:** The contractor looks at your house holistically and talks about upgrade options in terms of outcomes and benefits—it isn't a race to the bottom to get cheap equipment at the lowest price.

4. **Environmental control:** The contractor uses right-sized HVAC as a tool to solve problems that matter to homeowners, including thermal comfort and improving indoor air quality through filtration, ventilation, and humidity control.

5. **Necessary infrastructure:** The contractor confirms that the proposed system fits within the existing infrastructure in the house, including electrical panel and ductwork capacity.

I'm sold, but I need help.

Here are two more resources to help on your journey:

1. To see our constantly expanding library of tools for homeowners, visit www.foundryheatpumps.ca/resources

2. Register for a free HAVEN method webinar at www.foundryheatpumps.ca/webinars

If you're trying to find a great contractor in your area, reach out and I'll endeavour to connect you with a great contractor from my network.

Spark a feel-good community

Comfort and clean air lead to improved mental and physical health, increased productivity, fewer sick days, and reduced risk of respiratory issues and diseases.

Feel-good homes put people in the best environment to succeed. Imagine the compounding benefits across your friends and neighbours, with everyone enjoying better health, safer homes, and improved climate resilience.

That lifestyle should be available to everyone.

People are more likely to change their behaviors when they see friends and neighbours making changes.[49] It's a social contagion to conform to norms. Think about the campaign to stop smoking. We were presented with public health education campaigns around the health impacts of smoking, but the real tipping point was seeing others—especially high-profile, "cool" celebrities—reject smoking. There was a paradigm shift in public opinion, and the amount of smoking in public plummeted, for the betterment of public health and IAQ.

CONCLUSION: YOUR FEEL-GOOD HOME

The idea that houses can be havens of remarkable comfort, health, and sustainability can benefit from a similar shift. You can be the start of that change. Your next steps could be the spark that leads to a feel-good *community*.

Feel-good community

Live in a feel-good home

You did it! Finishing this book was the first step, and it's a step that most homeowners won't take. You've learned about the importance of valuing results and outcomes over upfront costs, the power of right-sized heat pumps, and you're armed with the understanding of good versus bad contractors.

I trust you linked the lessons with your own personal circumstances—I hope you are already thinking about ways to transform your house into a feel-good home. This book is a tool. It was designed to put you on the right path. The actual path, the right steps to take, and the ideal house all depend on you. You might need to read this book again, or you might want to review the sections that most inspired you.

Once you get there, oh boy—a feel-good home is the best version of your house. It's lavish in its comfort, it helps you coast through extreme weather and severe storms in a changing climate, and the clean air is a fresh start to a healthier life. Enjoy!

Notes

1 A Bailes, "What percent of time do you spend indoors?" (Energy Vanguard, August 31, 2018), www.energyvanguard.com/blog/what-percent-time-do-you-spend-indoors, accessed October 8, 2024
2 "Sick building syndrome" (United States Environmental Protection Agency, February 1991), www.epa.gov/sites/default/files/2014-08/documents/sick_building_factsheet.pdf, accessed November 10, 2024
3 B Orr, "How a heat pump reversing valve works" (HVAC School, February 12, 2021), https://hvacrschool.com/how-heat-pump-reversing-valve-works, accessed November 10, 2024

4 M Holladay, "Who can perform my load calculations?" (GreenBuildingAdvisor, March 24, 2017), www.greenbuildingadvisor.com/article/who-can-perform-my-load-calculations, accessed November 19, 2024

5 "Air sealing" (Energy Smart Home Performance, no date), https://energysmartohio.com/how_it_works/air-sealing, accessed November 13, 2024

6 A Bailes, "The surprising building science history behind the revolving door" (Energy Vanguard, January 29, 2016), www.energyvanguard.com/blog/the-surprising-building-science-history-behind-the-revolving-door, accessed January 14, 2025

7 MJ Mendell et al, "Carbon dioxide guidelines for indoor air quality: A review," *Journal of Exposure Science and Environmental Epidemiology*, 34/4 (2024), 555–569, https://doi.org/10.1038/s41370-024-00694-7

8 A Haddrell et al, "Ambient carbon dioxide concentration correlates with SARS-CoV-2 aerostability and infection risk," *Nature Communications*, 15/1 (2024), 3487, https://doi.org/10.1038/s41467-024-47777-5

9   H Canada, *Guidance for Fine Particulate Matter (PM2.5) in Residential Indoor Air* (Her Majesty the Queen in Right of Canada, represented by the Minister of Health, 2012), www.canada.ca/en/health-canada/services/publications/healthy-living/guidance-fine-particulate-matter-pm2-5-residential-indoor-air.html

10 R Ni et al, "Long-term exposure to PM2.5 has significant adverse effects on childhood and adult asthma: A global meta-analysis and health impact assessment," *One Earth*, 7 (2024), https://doi.org/10.1016/j.oneear.2024.09.022

11 J Lstiburek, "BSI-070: First deal with the manure and then don't suck" (BuildingScience.com, June 11, 2014), https://buildingscience.com/documents/insights/bsi-070-first-deal-with-the-manure, accessed November 11, 2024

12 A Bailes, "2 reasons to avoid most electronic air cleaners" (Energy Vanguard, April 23, 2021), www.energyvanguard.com/blog/2-reasons-to-avoid-most-electronic-air-cleaners, accessed November 13, 2024

13 J Rosenthal, "Could Corsi-Rosenthal boxes reduce particles to 'cleanroom' levels?" (Tex-Air Filters, September 2, 2023), www.texairfilters.com/could-corsi-rosenthal-boxes-reduce-particles-to-cleanroom-levels, accessed November 13, 2024

14 D Tozer, "The CLEAN Framework for evaluating air filters" (The Corsi-Rosenthal Foundation, July 22, 2024), https://corsirosenthalfoundation.org/articles/the-clean-framework-for-evaluating-air-filters, accessed November 13, 2024

15 A Bailes, "Humidity, Health, and Cold Climates" (GreenBuildingAdvisor, May 5, 2022), www.greenbuildingadvisor.com/article/humidity-health-and-cold-climates, accessed November 11, 2024

16 D Priestley, "The money making expert: The exact formula for turning $100 into $100k per month!" (The Diary of a CEO, February 22, 2024), www.youtube.com/watch?v=u0o3IlsEQbI, accessed November 11, 2024

17 EA Avallone, *Marks' Standard Handbook for Mechanical Engineers* [electronic resource] (Mcgraw-Hill, 2007)

18 "Toronto 1994 past weather (Ontario, Canada)" (2024), https://weatherspark.com/h/y/19863/1994/Historical-Weather-during-1994-in-Toronto-Ontario-Canada#Figures-Snowfall, accessed January 14, 2025

19 The Editors of Encyclopaedia Britannica, "Fahrenheit temperature scale: Definition, formula, and facts" (Britannica, no date), www.britannica.com/science/Fahrenheit-temperature-scale, accessed January 14, 2025

20 The Editors of Encyclopaedia Britannica, "Kelvin: Definition and facts" (Britannica, no date), www.britannica.com/science/kelvin, accessed January 14, 2025

21 "The Canada Green Buildings Strategy" (Natural Resources Canada, July 2022), https://natural-resources.canada.ca/sites/nrcan/files/engagements/green-building-strategy/CGBS%20Discussion%20Paper%20-%20EN.pdf, accessed May 6, 2024

22 D Khojasteh et al, "Climate change and COVID-19: Interdisciplinary perspectives from two global crises," *The Science of the Total Environment*, 844 (2022), 157142, https://doi.org/10.1016/j.scitotenv.2022.157142

23 B Bow et al, "Acute and postacute sequelae associated with SARS-CoV-2 reinfection," *Nature Medicine*, 28 (2022) 2398–2405, https://doi.org/10.1038/s41591-022-02051-3

24 SI Seneviratne et al, "Weather and climate extreme events in a changing climate," in: *Climate Change 2021—The Physical Science Basis Working Group: Contribution to the Sixth Assessment Report of the Intergovernmental Panel on Climate Change* (Cambridge University Press, 2021), pp1513–1766, https://doi.org/10.1017/9781009157896.013, accessed January 14, 2025

25 C Dalton, "Why heat waves of the future may be even deadlier than feared," *The New York Times* (October 26, 2024), www.nytimes.com/2024/10/25/health/heat-tolerance-climate-change.html, accessed November 12, 2024

26 P Stott, "How climate change affects extreme weather events: Research can increasingly determine the contribution of climate change to extreme events such as droughts," *Science*, 352/6293 (2016),

1517–1518, https://doi.org/10.1126/science.aaf7271
27 E Uguen-Csenge and B Lindsay, "For 3rd straight day, B.C. village smashes record for highest Canadian temperature at 49.6 C" (CBC, June 30, 2021), www.cbc.ca/news/canada/british-columbia/bc-alberta-heat-wave-heat-dome-temperature-records-1.6084203, accessed November 12, 2024
28 A Bailes, "The 4 factors of comfort" (Energy Vanguard, June 7, 2010), www.energyvanguard.com/blog/the-4-factors-of-comfort, accessed January 14, 2025
29 A Bailes, "We are the 99%—design temperatures and oversized HVAC systems" (Energy Vanguard, May 10, 2022), www.energyvanguard.com/blog/we-are-the-99-design-temperatures-oversized-hvac-systems, accessed November 12, 2024
30 "ASHRAE climatic design conditions 2009/2013/2017/2021" (Ashrae-Meteo.info), https://ashrae-meteo.info/v2.0, accessed November 12, 2024
31 K Saleeby, "Mean radiant temperature: What it is and why we should care"

(HVAC School, May 22, 2021), https://hvacrschool.com/mean-radiant-temperature-what-it-is-and-why-we-should-care, accessed November 11, 2024

32 A Bailes, "Manual J load calculations vs. rules of thumb" (GreenBuildingAdvisor, September 7, 2016), www.greenbuildingadvisor.com/article/manual-j-load-calculations-vs-rules-of-thumb, accessed November 13, 2024

33 D Dorsett, "Replacing a furnace or boiler" (GreenBuildingAdvisor, April 18, 2016), www.greenbuildingadvisor.com/article/replacing-a-furnace-or-boiler, accessed November 13, 2024

34 N Adams, *The Home Comfort Book: The ultimate guide to creating a comfortable, healthy, long lasting, and efficient home* (Createspace Independent Publishing Platform, 2017)

35 J Lstiburek, "BSI-053: Just right and airtight" (BuildingScience.com, September 15, 2011), https://buildingscience.com/documents/insights/bsi-053-just-right-and-airtight, accessed November 13, 2024

36 A Fremeth and G Holburn, *A Historical and Comparative Perspective on Ontario's Electricity Rates Policy Brief* (The Ivey

Energy Policy and Management Centre, 2018), www.ivey.uwo.ca/media/3782393/july-2018-a-historical-and-comparative-perspective-on-ontario-s-electricity-rates.pdf, accessed January 14, 2025

37 S Evans, "Solar is now 'cheapest electricity in history,' confirms IEA," *CarbonBrief* (October 13, 2020), www.carbonbrief.org/solar-is-now-cheapest-electricity-in-history-confirms-iea, accessed January 14, 2025

38 D Schlissel and D Wamsted, "Small modular reactors: Still too expensive, too slow and too risky" (Institute for Energy Economics and Financial Analysis, May 29, 2024), https://ieefa.org/resources/small-modular-reactors-still-too-expensive-too-slow-and-too-risky, accessed November 14, 2024

39 A Bailes, "Does a geothermal heat pump count as a renewable energy source?" (Energy Vanguard, July 12, 2012), www.energyvanguard.com/blog/does-a-geothermal-heat-pump-count-as-a-renewable-energy-source, accessed November 14, 2024

40 A Bailes, "Efficient hot water delivery with a simple tool" (Energy Vanguard,

May 26, 2022), www.energyvanguard.com/blog/efficient-hot-water-delivery-with-a-simple-tool, accessed November 14, 2024

41 S Gibson, "Drain water heat recovery gets a boost in Ontario," *GreenBuildingAdvisor* (April 18, 2023), www.greenbuildingadvisor.com/article/drain-water-heat-recovery-gets-a-boost-in-ontario, accessed November 14, 2024

42 W Armand, "Have a gas stove? How to reduce pollution that may harm health" (Harvard Health Publishing, September 7, 2022), www.health.harvard.edu/blog/have-a-gas-stove-how-to-reduce-pollution-that-may-harm-health-202209072811, accessed November 14, 2024

43 ED Lebel et al, "Methane and NOx emissions from natural gas stoves, cooktops, and ovens in residential homes," *Environmental Science and Technology*, 56/4 (2022), https://doi.org/10.1021/acs.est.1c04707

44 D Roberts, "Gas stoves can generate unsafe levels of indoor air pollution," *Vox*

(May 11, 2020), www.vox.com/energy-and-environment/2020/5/7/21247602/gas-stove-cooking-indoor-air-pollution-health-risks, accessed November 14, 2024

45 "Ford Model T Specs, Dimensions" (Ultimate Specs, n.d.), www.ultimatespecs.com/car-specs/Ford/20867/Ford-Model-T-.html, accessed February 10, 2025

46 R Flatter, "Secretariat remains No. 1 name in racing" (ESPN, 2019), www.espn.com/sportscentury/features/00016464.html, accessed January 14, 2025

47 C Woodford, "How do induction cooktops work?" (Explain That Stuff, January 6, 2019), www.explainthatstuff.com/induction-cooktops.html, accessed January 14, 2025

48 A Biro, "What is passive house?" *gb&d Magazine* (May 9, 2024), https://gbdmagazine.com/what-is-passive-house, accessed November 14, 2024

49 M Bergquist et al, "Field interventions for climate change mitigation behaviors: A second-order meta-analysis," *Proceedings of the National Academy of Sciences*, 120/13 (2023), https://doi.org/10.1073/pnas.2214851120

Acknowledgments

A special thank you to Nate Adams for his constant support and foundational knowledge, and for selflessly sharing thought leadership in the industry.

To the Jouleia team—Paul Sehr, Patrick Marshall, and Mat Krizmanich—for stoking an entrepreneurial spirit that led me down this path.

To Travis Richardson, for giving me the opportunity and professional space to explore the ideas in this book.

To Laura Tozer and Adam Scott, for acting as guiding lights in my career.

To my mom and dad, Catharine Tozer and Glenn Tozer, for my love of writing and numbers, respectively.

Thank you.

The Author

Drew Tozer gained expertise in building science and heat pumps as an NRCan-registered energy advisor before partnering with a local HVAC company and rebranding as Foundry Heat Pumps. Prior experience included ten years in the public and private sectors for renewables and energy conservation.

He's passionate about building science, heat pumps, and climate change.

He runs business development for Foundry Heat Pumps near Toronto, Ontario, and contributes to expert advisory panels, industry podcasts, and traditional media.

Drew Tozer is a subject matter expert in the industry and continues to be a strong advocate for better homeowner experiences by pushing for right-sized HVAC as a tool to solve problems that matter to homeowners.

He knows that electrification is an important step in the fight against climate change, and right-sized heat pumps are a path for homeowners to reduce emissions while improving their quality of life.

- www.foundryheatpumps.ca
- www.linkedin.com/in/drewtozer
- @drewtozer.bsky.social